Statistics in Language Research: Analysis of Variance

Statistics in Language Research: Analysis of Variance

by

Toni Rietveld and Roeland van Hout

Mouton de Gruyter
Berlin · New York

Mouton de Gruyter (formerly Mouton, The Hague)
is a Division of Walter de Gruyter GmbH & Co. KG, Berlin.

♾ Printed on acid-free paper which falls within the guidelines of the ANSI to ensure permanence and durability.

Library of Congress Cataloging-in-Publication Data

Rietveld, Toni, 1949–
Statistics in language research : analysis of variance / by Toni Rietveld and Roeland van Hout.
p. cm.
Includes bibliographical references and index.
ISBN-13: 978-3-11-018580-5 (cloth : alk. paper)
ISBN-10: 3-11-018580-6 (cloth : alk. paper)
ISBN-13: 978-3-11-018581-2 (pbk. : alk. paper)
ISBN-10: 3-11-018581-4 (pbk. : alk. paper)
1. Linguistics – Statistical methods. 2. Analysis of variance. I. Hout, Roeland van. II. Title.
P138.5.R543 2005
410'.2'1–dc22
2005019159

Bibliographic information published by Die Deutsche Bibliothek

Die Deutsche Bibliothek lists this publication in the Deutsche Nationalbibliografie; detailed bibliographic data is available in the Internet at <http://dnb.ddb.de>.

ISBN-13: 978-3-11-018580-5 hc.
ISBN-10: 3-11-018580-6 hc.
ISBN-13: 978-3-11-018581-2 pb.
ISBN-10: 3-11-018581-4 pb.

Cover design: Klein Kon Jung, Berlin
Printed in Germany.

Contents

Acknowledgements

We are indebted to our students for their comments on earlier drafts of the book. We thank dr. Han Oud of Radboud University of Nijmegen for the hours he was so kind to spend considering the pros and cons of maximum likelihood estimation in mixed models. We thank Lieselotte Toelle MA for correcting our English, Dr. Bert Cranen and drs. Joop Kerkhoff for the art-work. Needless to say, we are the ones responsible for any remaining errors.

Chapter 1
Language research and statistics

1.1. Statistics and analysis of variance in language research

Language research is based on data. Sometimes the data are quite subjective, like the appreciation of voices or accents, or introspective, like the intuitions of linguists on the well-formedness of utterances. In many situations an introspective approach is warranted and one does not need quantitative or statistical methods to corroborate the scientific argumentation. However there are many situations in which the language researcher needs to collect data through a survey, in a field study, in an experiment, or in a language corpus. Many linguistic subdisciplines use methods which are similar to the ways researchers in the social sciences obtain and analyze data.

Researchers in these subdisciplines want to generalize the outcomes to the population(s) from which they have taken one or more samples. The wish to make generalizations entails the use of inductive statistics, the branch of statistics which enables us to infer population characteristics. An example: "Speakers in dialect A exhibit phonetic process X more often than speakers of dialect B", on the basis of outcomes obtained in relatively small samples, for instance 110 speakers of dialect A and 80 speakers of dialect B. As is the case in psychology and sociology, sampling a population in such a way that the sample is representative of the population under consideration in gender, age, socio-economic status etc. is an important issue. Random sampling is a skill in itself, especially in survey research.

This book does not deal with data collection but with data analysis, and, more particularly, it deals with ANalysis Of VAriance, abbreviated ANOVA. This technique is the main instrument for social scientists and their linguistic colleagues to analyze the outcomes of research designs with more than two treatments or groups. Moreover, analysis of variance enables the researcher to assess the effects of more than one independent variable at the same time. When data are obtained from participants of two different groups at four points in time, we may want to know whether the outcomes of the two groups differ; we may also want to know whether their outcomes change at different rates.

Often students and researchers need to apply analysis of variance to their data although they may feel insecure about the basic principles of statistical testing. That is why in this book the treatment of analysis of variance is preceded by two chapters which explain the use of t tests, type I and type II errors and power analysis. These are fundamental concepts which constitute the basis of statistical testing. Special attention is paid to an important concept in experimental design: determining the size of the sample needed to detect specific hypothesized effects in (the) population(s).

This book gives a comprehensive treatment of ANOVA. Technical and more complete treatments can be found in Kirk (1995) and Winer, Brown, and Michels (1991), but these two textbooks are fairly hard to comprehend for those researchers in the field of language and speech behaviour who have only attended an introductory course in statistics, perhaps based on specific textbooks for language research like Butler (1985) and Woods, Fletcher, and Hughes (1986). Many language researchers seem to use ANOVA simply by following the HELP files of statistical packages like SPSS or books which are often obsolete as far as current developments are concerned. We want to explain to language researchers what they are doing when they use ANOVA and which options are available. In addition, we would like to inform them about developments in post hoc comparisons, power analysis, standards in reporting statistics, ways of dealing with missing data, and the pros and cons of the F1 × F2 approach currently used in psycholinguistic research. Another problem with books on ANOVA is that they only deal with examples from the social sciences and are often either too simple or too complicated for the user of statistics.

In Chapters 4 to 9 we cover the most widely used experimental designs step by step, showing the researcher how the analyses have to be executed. Chapters 9 and 10 are slightly more complicated than the others we admit, but we wanted to highlight more recent developments in the analysis of multi-group data: different estimation procedures, multilevel analysis, bootstrap and randomization tests. A specific section in Chapter 9 deals with missing data, a common phenomenon in psycho- and sociolinguistic research, which is more relevant than most researchers assume: The standard approach in reporting analysis of variance is not to deal with missing data.

The structure of a chapter is determined by the statistical concepts and analyses handled, but the following sections return in each chapter:

- Preview: Information is provided about what one is going to read.

- Technical sections with examples taken from linguistic research.
- Terms and concepts: Summary of the concepts presented in the chapter.
- Statistical reporting: Examples are presented, mostly based on suggestions made by the American Psychological Association (APA).
- Exercises.

1.2. Variables

There are a number of data collection methods in linguistics, such as survey research, experimental research and corpus research. In all these methods the *variable* concept plays a key role. A variable is a property or characteristic of a person, a condition, an object, or any other research element. These are defined by the research questions and the way in which they are made operational in the research procedures. In the examples we use in this book, the elements are often subjects, informants, language users or dialect speakers. We usually call them *participants*, to meet the more recent standard terminology in reporting research.

Often we want to know whether variable A affects scores obtained on variable B. We call variable A an independent variable, and variable B a dependent variable, a distinction which is particularly familiar in (quasi-)experimental research. The independent variable is not always under the control of the researcher. If, for instance, speakers of dialect A_1 live in rural areas, whereas speakers of dialect A_2 are mainly found in urban areas, the researcher cannot change this fact. Being a speaker of dialect A_2 is connected to being an urbanite. In genuinely experimental research the investigator has the independent variable(s) under full control. He/she can deliberately introduce four levels of noise (50, 60, 70 and 80 dB) in which participants have to identify specific speech segments, or specify the number of syllables (1, 2, 3) of carrier words which participants have to listen to. In the examples discussed there is at least one dependent and one independent variable.

Variables have values which they get as a result of a measurement procedure. Measurement is the assignment of numerals to objects or events according to rules, cf. Stevens (1946). The scale of measurement determines the amount of information contained in the data (Anderson, Sweeney, and Williams 1991: 19). There are four scales of measurement:

1. *Nominal scale* (also called 'categorical scale' or 'qualitative' scale). Language background is a good example. One cannot say that English

is a better or more sophisticated language background than Dutch. It is simply a different language, the way an adjective is different from a noun. A variable is nominal if it is used to label elements or observations in order to categorize or classify them. All transformations are allowed which leave classification unaffected: A, B and C can be transformed into $\gamma, \delta, \varepsilon$, into the numbers 1, 2 and 3, or the other way round, into 3, 2 and 1, and even into –5, 11236 and 432 etc. as long as the different labels represent different classes. Objects measured on nominal scales can only be distinguished from each other. They are equal or not equal, i.e. they belong to the same class or category (= equal) or not (= not equal); object K = object M or object K $\neq$ object M.

2. *Ordinal scale.* In addition to distinguishing objects (on the nominal scale), we can also rank objects which are measured on an ordinal scale (either object K > object M or object M > object K). Language competence is an example of an ordinal scale. Advanced learners of English know more about English than intermediate learners and the latter know more than beginners. Students with these labels differ in their competence of English, and can be ordered on the variable 'competence of English'. More information is available than information measured on a nominal scale. We do not know, however, whether the difference between advanced and intermediate students is equal to the difference between an intermediate and a beginner. Theoretically all monotone transformations are permitted, that is all transformation which leave the order unaffected. Strictly speaking a monotone transformation of scale values like 1, 2, 3, 4 (representing, for instance, degrees of harshness) into 1, 3, 37, 58 is permitted as the order is not affected. In practice, such a transformation is disturbing, for the numbers suggest that we do have more precise information. On an ordinal scale the intervals do not contain any information.
3. *Interval scale.* If the difference between measurement levels is known, the data are measured on an interval scale. For instance, the physical difference between the two pitch values 100 and 120 Hz is equal to the difference between 200 and 220 Hz (viz. 20 Hz). It is another question whether the same holds for perceived pitch differences. Linear transformations of the $F(X) = aX + b$ type are permitted as they leave the differences between scale values unaffected.
4. *Ratio scale.* An absolute requirement for a ratio scale is that a true zero

value is defined on the scale. A car which does not cost anything ('is for free') gets the value '0' on this scale. Ratios make sense: 400 kilos are twice as heavy as 200 kilos, and 200 kilos are twice as heavy as 100 kilos. This is not the case for data measured on an interval scale. We cannot say that the coronation of Charlemagne (800 AD in Aix-la-Chapelle) was twice as late as the 'Great Migration' (400 AD). We have a zero value in our calendar time, but it is an arbitrary zero value. The transformation $F(X) = aX$ is permitted here. Although it changes the absolute value of the difference, it does not change the relative value. In contrast to the interval scale adding b to the transformation is not permitted, as it would affect the absolute zero value.

The scale level on which variables are measured has consequences for the statistical technique that can be applied. It is customary to say that t tests and analysis of variance require strict interval data for the dependent variable. We do not think this is the case. The robustness of these techniques is amazing, but we postpone the discussion to Chapter 7.

A concrete example may illustrate the distinctions we make here. Let us assume that we have a nominal variable 'region'. If a researcher wants to know whether the nominal variable 'region' affects the duration of a vowel associated with a sentence accent (measured in milliseconds), we have the independent variable 'region' and the dependent variable 'duration'. Each measurement carried out on speakers from the region in question, constitutes a 'case'. In the following matrix we display nine *cases* rowwise, and two variables columnwise. The first column contains the values of the dependent variable, the second one the values of the independent variable.

Table 1.1. A data matrix with nine cases and two variables

case	region	duration
1	1	120.000
2	1	124.000
3	1	130.000
4	2	130.000
5	2	140.000
6	2	145.000
7	3	120.000
8	3	110.000
9	3	100.000

Most statistical packages have a standard data format: Cases are represented row-wise, variables column-wise. Cases and variables are defined by the research design, and a great variety of cases and variables is possible. In the examples we present in our book we restrict ourselves to straightforward research designs, where participants play the roles of cases. The data in Table 1.1 show this data format. The independent variable 'region' is nominal and distinguishes three regions. The dependent variable 'duration' is a ratio variable with a true zero point. There are three cases per region.

Finally, we would like to say a few words about corpus research, which has become very popular in the last two decades. The fact that large databases have become available is an important factor in the increasing use of corpora as sources of information, next to intelligent research tools. Very often the outcome variables in this kind of research consist of 'counts' and relative frequencies, often converted to percentages, like "X% of the sentences are shorter than 10 words in text type A, whereas it is Y% in text type B". Most of the time analysis of variance is not the appropriate tool for the analysis of this kind of data. We refer to Baayen (2001) for word frequency statistics, and to Oakes (1998) for statistics in corpus linguistics. For pitfalls in corpus research see Rietveld, Van Hout, and Ernestus (2004).

1.3. Designs

In this book we are going to deal with the statistical analysis of data obtained in a number of so-called research *designs*, which define the research variables and their status. Two relevant questions for the characterization of a design with one dependent and one independent variable are the following:

- *How many levels or values has the independent variable?* This question relates to the number of classes or categories distinguished in the nominal independent variable. If two levels are distinguished a t test can be applied. With more than two classes an ANOVA technique is required.
- *Is the independent variable a between-subject factor or a within-subject factor?* In studying language acquisition we can track the development over time by using different age groups, each group consisting of different children. Suppose we test them with a vocabulary test. By comparing the age groups we investigate vocabulary development; the differences we observe are differences between children in different age groups. So development is a between-participant factor, or in the classical terminol-

ogy we use, a between-subject factor. It concerns differences between groups. The other route is to track the development of a group of children over time. The same group is tested repeatedly over a longer period of time. Such a design is called longitudinal. The differences we observe are differences within the same children. Development now is a within-participant factor, or in the classical terminology, a within-subject factor.

The distinctions related to these two questions were taken into account in the schedule in Table 1.2, where crossing the two questions delivers four basic designs.

Table 1.2. Four basic designs based on the properties of the independent variable

independent variable	two levels	more than two levels
between-subject factor	*design 1* *t* test independent samples	*design 3* one-way ANOVA
within-subject factor	*design 2* *t* test correlated samples	*design 4* repeated measures ANOVA

On the next pages we present schedules of the four basic designs. They are represented in the SPSS data matrix format. Research design 1 comprises two groups (equal numbers of cases are not required). In experimental conditions, the participants are assigned to the two groups in an a-select way. Table 1.3 distinguishes one independent variable, i.e. 'group' with two levels. The dependent variable is called 'dep'.

We did not assign each case (participant) a unique case label. In practice, we should do so, to be sure that the right data are assigned to the right participant. In SPSS a case automatically gets a case label. We assigned the levels 1 and 2 to the 'group' variable, but, as said above, the real values are not relevant as long as they are different. The values –103 and 3425 yield the same outcomes in the statistical analysis.

Design 2 is represented in Table 1.4. In this design we have a within-subject factor or variable, which means that the same participants are measured twice in two different situations or conditions. In fact we have one group of participants.

The data format in Table 1.4 looks completely different from the format in Table 1.3. In both tables, each case (= participant) is represented by one

Table 1.3. Design 1: SPSS data matrix, the independent between-subject factor or variable 'group' with two levels

group	dep
1	15
1	16
1	28
1	32
1	12
2	20
2	32
2	16
2	11
2	14

Table 1.4. Design 2: SPSS data matrix, the independent within-subject factor is represented by two variables

cond1	cond2
15	19
16	33
28	28
32	44
12	18

row. Participants in design 2 were measured twice, which is represented by two variables, the first one containing the scores obtained in the first condition 'cond1', and the second one the scores obtained in the second condition, 'cond2'. The constraint of having only two levels for the independent variable is expressed by just having two variables to represent the 'condition' factor.

Design 3 is an expansion of design 1 for the number of levels of the 'group' variable, which now distinguishes three values or levels. This can be seen in Table 1.5, which does not contain values for the dependent variable.

Design 4 is an expansion of design 2. Instead of two variables we now have three, which means that the participants were measured three times. The three measurements constitute the within-subject factor. Such designs are often labelled as repeated measures designs. The SPSS data matrix format is given in Table 1.6.

Table 1.5. Design 3: SPSS data matrix, the independent between-subject variable 'group' has three levels

group	dep
1	
1	
1	
1	
1	
2	
2	
2	
2	
3	
3	
3	
3	

Table 1.6. Design 4: SPSS data matrix, the independent within-subject factor 'time' is represented by three variables, 't1', 't2', and 't3'

t1	t2	t3

The subsequent step to expand the analytical possibilities is to increase the number of factors. Starting from the single between-subject and within-subject factor designs, we can add three designs:

- design 5: a multifactorial design for between-subject factors
- design 6: a multifactorial design for within-subject factors
- design 7: a multifactorial mix of between- and within-subject factors.

Design 5 is a so-called *completely randomized factorial design.* The two independent variables could be gender (2 levels) and experimental condition

(3 levels). Unequal numbers of participants often lead to problems, as we discuss later (Chapter 7). An example with the variables gender and experimental condition is given in Table 1.7.

Table 1.7. Design 5, SPSS data matrix, completely randomized factorial design with two independent variables 'gender' and 'cond' (= condition)

cond	gender	dep
1	1	
1	1	
1	2	
1	2	
2	1	
2	1	
2	2	
2	2	
3	1	
3	1	
3	2	
3	2	

The variable 'dep' in Table 1.7 is the dependent variable. One may add other independent variables, but the number of combinations will then quickly multiply. Adding a third facor with three levels results in 18 combinations. A large number of cells impedes the interpretation of the statistical outcomes.

Design 6 is a multifactorial within-subject design. All participants are measured in all factor combinations, as is shown in Table 1.8. There are two within-subject factors, 'time' (three levels) and 'condition' (two levels) which means 6 combinations. Consequently, there are six variables in Table 1.8. The subscript of factor 'c' changes more slowly than that of factor 't', moving from left to right.

Finally, we have design 7, the so-called *split-plot design*, with at least one within-subject factor and one between-subject factor. The latter refers to the independent variable which distinguishes participants, like gender or socio-economic status. In our example in Table 1.9 there is one between-subject factor 'therapy' with two levels (two kinds of therapy) and one within-subject factor, again 'time' with 3 levels: 't1', 't2', 't3'.

Table 1.8. Design 6: SPSS data matrix, with two independent within-subject factors 'time' (three levels, indicated by 't1', 't2', 't3') and 'condition' (two levels, indicated by 'c1', 'c2'), represented by six variables

c1t1	c1t2	c1t3	c2t1	c2t2	c2t3

Table 1.9. Design 7: SPSS data matrix, with a between-subject factor 'therapy' and an independent within-subject factor 'time' represented by three variables, 't1', 't2', and 't3'

therapy	t1	t2	t3
1			
1			
1			
1			
1			
1			
2			
2			
2			
2			
2			
2			

The three time variables in Table 1.9 could represent a pre-test, a post-test and a post-test after a longer period of time, to check whether the effect of a therapy remains or fades.

Other aspects can complicate the designs we have discussed. In the multifactorial designs presented above, all combinations of levels of all factors occur. The resulting design is called a crossed design. However, there are hierarchical designs as well, these are designs in which not all levels of one factor co-occur within all levels of another factor. An example is the factor denomination of hospitals (Christian and General) and the Hospitals themselves (3 Hospitals of each denomination). Each hospital is only listed in one denomination. We will discuss nesting in Chapter 5.

A final question here is whether the values of the independent variable represent a sample. Do the independent variables (factors) involve random samples of a large number of possible samples? If the answer is yes, we have to deal with a so-called random factor, if not, the factor is called a fixed factor. This distinction affects the way in which the data have to be analyzed, as we discuss in Chapter 5.

1.4. Statistical packages

Hardly anyone carries out statistical analyses by hand or by pocket calculator. It became the task of statistical computer packages, supplemented by dedicated software for less common statistical procedures. We mention SAS: Statistical Analysis System, SPSS : Statistical Package for the Social Sciences, MINITAB, S+, R. We do not want to express a preference for one of these, but we chose SPSS (version 12.0) to provide examples of analysis in this book, as we think it is the most widely known package in language research.

We will demonstrate how statistical analyses can be carried out in SPSS with the Point-And-Click (PAC) window system. We show how to use SPSS syntax in a number of cases, because the syntax approach offers extra options and possibilities in carrying out a statistical analysis.

Chapter 2
Basic statistical procedures: one sample

2.1. Preview

In this chapter we review the basic principles of statistical testing. These principles are reviewed on the basis of the *one-sample* design with the dependent variable measured at the interval or ratio level. A one-sample design is a design in which a statistic of a sample drawn from a population, for instance a mean value, is compared with a value which is hypothesized to hold for the population. Obviously, a sample mean will hardly ever have the same value as the hypothesized value ($\bar{X}$ is not μ): If we hypothesize that the mean lifetime of lightbulbs is 2000 hrs, and we draw a random sample of 100 bulbs, the mean lifetime of that sample cannot be expected to be 2000 hrs, but a figure just above or below this, even if the mean liftetime in the population (μ) is 2000 hrs. The question is what degree of deviation of the observed value of the mean from the hypothesized value can be accepted without us having doubts on the hypothesized mean value. The concept of *sampling variability* is extremely important in this context (Section 2.2). In Section 2.3 we review the procedure of hypothesis testing and in Section 2.4 the well-known *t* distribution. Important concepts are *statistical power* and *effect size* (Section 2.5). In Section 2.6 we discuss the calculation of the *sample size needed* to detect a pre-defined effect size. These concepts tend to be (wrongly) neglected in linguistic research, but they are very important in medical research. An example is the test of a hypothesis that phoneticians only guess when they are asked to assess the age of a speaker, against the hypothesis that their ratings are correct in, for example, 75% of the cases.

2.2. Sampling variability

With continuous (interval or ratio) data, our interest is often focused on the mean(s) of our sample(s). However, the mean only provides a summary of the data we have collected. If we were to collect another sample of the same size, it is extremely unlikely that precisely the same mean value will be obtained. This phenomenon is called *sampling variability*.

Let us look at a simple situation, the reaction time for recognizing a Dutch word. The time taken to recognize the word 'rente' (= interest) was measured. The mean time taken to recognize the word was 523 ms. This is the mean of our sample of Dutch listeners, but we would like to be able to generalize this to all native listeners of Dutch, not just the ones we happen to have sampled. Our best guess of the mean time taken to recognise 'rente' in the population is the mean of our sample, 523 ms. This is all the information available, apart from the standard deviation, and we have no reason to expect it to be biased in any way.

Of course, we would like to know whether the value of 523 ms – denoted by $\bar{X}$ – is a good estimate of the population mean μ. In order to get an intuitive feeling of the quality of our estimate, we have to know more about the sampling variability. To that end we introduce the concepts of *sampling distribution* and *standard error*, *SE*. If we take a large number of samples of, for instance 30 observations, and calculate the mean of each sample, the sample means will be normally distributed around μ. Even if the distribution is skewed or uniform, the distribution of sample means is normal. This important fact is called the *Central Limit Theorem*, a theorem which plays a crucial role in statistics. The distribution of sample means also has a standard deviation, called the *standard error*, which can be estimated from the data itself. If we can assume that the size of the sample is smaller than or equal to 5% of the population size, the formula for the standard error of the sample means, $\bar{X}$ is (cf. Anderson, Sweeny, and Williams 1991: 231):

$$\boxed{SE_{\bar{X}} \quad = \quad \sigma_{\bar{X}} = \frac{\sigma}{\sqrt{n}} \qquad (1)}$$

where σ is the standard deviation of the population of our data, and n is the sample size. In most cases σ is unknown, and has to be estimated by the standard deviation of the sample: s. In our example, $s = 76.44$ and $n = 28$, thus the estimated standard error of the mean is $76.44/\sqrt{28} = 14.45$.

The formula for the standard error summarizes two factors which affect the stability of the estimates of the mean of a population μ: the variation in the population σ (estimated by s) and the size of the sample n. The larger the variation in the population is, the more we can expect sample means $\bar{X}$ to vary – that is why s is in the numerator of the formula; the larger n is, the smaller the fluctuations in sample mean values are.

The Central Limit Theorem tells us that a statistic like $\bar{X}$ is normally distributed with the mean μ and the standard deviation $\sigma/\sqrt{n}$ if the number of observations is large enough (in theory the theorem holds when n approaches infinity, in practice $n > 30$ suffices). Supposing we have a normally distributed population, with $\mu = 250$ and $\sigma = 50$, we can randomly draw (by compute) 500 samples of size $n = 3$, of $n = 30$, and of $n = 100$. The distributions of the 500 samples for the three samples sizes are given in Figure 2.1.

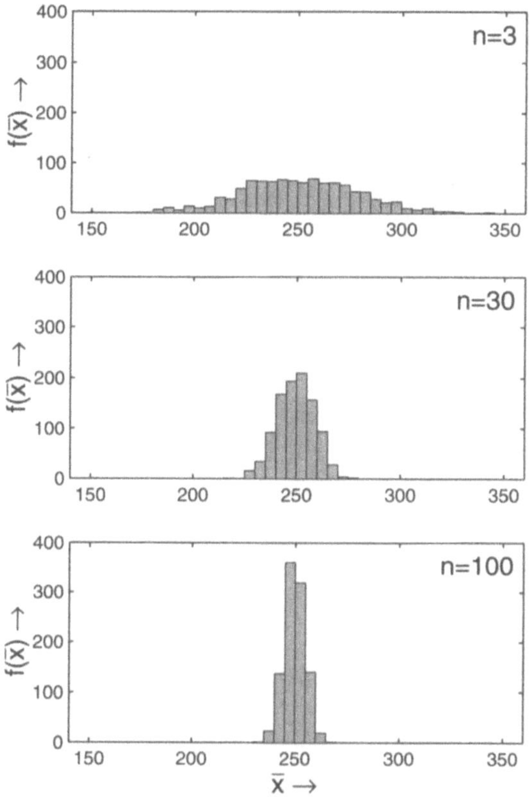

Figure 2.1. Distributions of 500 sample means for three different sample sizes: $n = 3, n = 30, n = 100$, with samples drawn from a normally distributed population with $\mu = 250$ and $\sigma = 50$

The *SE* of the sampling distribution with $n = 3$, is $50/\sqrt{3} = 28.90$; the *SE* for samples with $n = 30$ is $50/\sqrt{30} = 9.13$. For sample size $n = 100$, we get $50/\sqrt{100} = 5$. The effect of n becomes quite clear: The larger n is, the smaller the *SE* of the sampling distribution becomes. This is shown by the three sample distributions in Figure 2.1. At the same time, we see that the samples have a normal distribution, particularly with larger sample sizes.

Figure 2.1 shows that the standard error of the sample mean decreases if the sample size increases. Let us take a closer look at this relationship. Given the sample mean, the standard error and the sample size are connected by the square root function. Some examples are given in Table 2.1, in which a standard deviation of 100 is assumed in the population.

Table 2.1. Relationship between sample sizes and the standard error, given a standard deviation of 100

sample size	square root sample size	standard error
1	1	100
4	2	50
16	4	25
64	8	12.50
256	16	6.25
1024	32	3.125

Starting at the value 1 the sample sizes are quadrupled. Going from sample size 1 to 4, the standard error decreases with $1/\sqrt{4} = 1/2$, which entails a drop of the standard error from 100 to 50. Quadrupling the sample of 4 to 16 again returns a drop by 1/2. The conclusion is obvious. Within the range of smaller sample sizes, increasing the sample size has a large impact on the standard error. For larger samples, there is still an impact, but it is less pronounced.

2.3. Hypothesis testing: one sample

When an $\bar{X}$ – a sample mean – is available, we might like to know whether its value is in accordance with an expectation. Especially in industrial settings people might have expectations about the value of μ in the population, and confront these with the actual sample means (for instance of the mean lifetime of light bulbs). If the observed sample mean substantially deviates from the expectation, then one has to revise the expectation (or sue the manufacturer...). In this case the standard normal distribution – also called the z distribution – offers help. As you may remember from basic statistics, a transformation of a raw score into a z score yields a variable with 'standard properties'. The resulting variable has a mean of 0, and a standard deviation of 1. This is also written as $N(0, 1)$. The equation is:

$$z = \frac{X - \mu}{\sigma} \qquad (2)$$

We can also standardize statistics like $\bar{X}$ or the difference between two means: $\bar{X}_1 - \bar{X}_2$. The only thing we need is the appropriate term for the denominator: *SE*, which takes different forms for different statistics. For the statistic $\bar{X}$ the z value is:

$$z = \frac{\bar{X} - \mu}{\sigma/\sqrt{n}} \qquad (3)$$

You may ask what standardization is for? Well it enables us to use tables to find out whether the deviation of our sample mean from the hypothesized population mean is a probable one. Let us assume that we have a hypothesis about the mean μ of the reaction times mentioned above, for example 490 ms. The sample mean is 523 ms, the standard deviation 76.44, and the sample size 28. Filling in these values we get:

$$z = \frac{523 - 490}{76.44/\sqrt{28}} = 2.28 \qquad (4)$$

We just have to use the well-known tables of the standard normal distribution to find out that the probability of obtaining a z equal to or larger than 2.28 is 0.0113: $p(z \geq 2.28) = 0.0113$. Below we reproduce two sections of this kind of table which can be organized in two different ways:

1. The upper part (= a) in Table 2.2 shows the probability of obtaining z values between 0 and the observed z value (= Z) ($p(0 \leq z \leq Z)$) (see section a in Table 2.2); in order to obtain the probability of a z value equal to or larger than the actual Z value, we subtract the probability found from 0.50, as the probability surface left and right from $z = 0$ equals 0.50; cf. Figure 2.2, the distribution in the right panel.
2. The lower part (= b) in Table 2.2 shows the probability of obtaining z values equal to or larger than the actual Z value; cf. Figure 2.2, the distribution in the left panel.

Referring to our example, with $z = 2.28$, we obtain in part (a) of Table 2.2 at the intersection of $z = 2.2$ and 0.08: 0.4887; 0.5000 – 0.4887 = 0.013. In part (b) we get this result straight away: 0.013.

Table 2.2. Sections of tables of the standardized normal distribution

a $p(0 \leq z \leq Z)$

z	.00	.01	.02	.03	.04	.05	.06	.07	.08	.09
1.9	.4713	.4719	.4726	.4732	.4738	.4744	.4750	.4756	.4761	.4767
2.0	.4772	.4778	.4783	.4788	.4793	.4798	.4803	.4808	.4812	.4817
2.1	.4821	.4826	.4830	.4834	.4838	.4842	.4846	.4850	.4854	.4857
2.2	.4861	.4864	.4868	.4871	.4875	.4878	.4881	.4884	.4887	.4890

b $p(z \geq Z)$

z	.00	.01	.02	.03	.04	.05	.06	.07	.08	.09
1.9	.0287	.0281	.0274	.0268	.0262	.0256	.0250	.0244	.0239	.0233
2.0	.0228	.0222	.0217	.0212	.0207	.0202	.0197	.0192	.0188	.0183
2.1	.0179	.0174	.0170	.0166	.0162	.0158	.0154	.0150	.0146	.0143
2.2	.0139	.0136	.0132	.0129	.0125	.0122	.0119	.0116	.0113	.0110

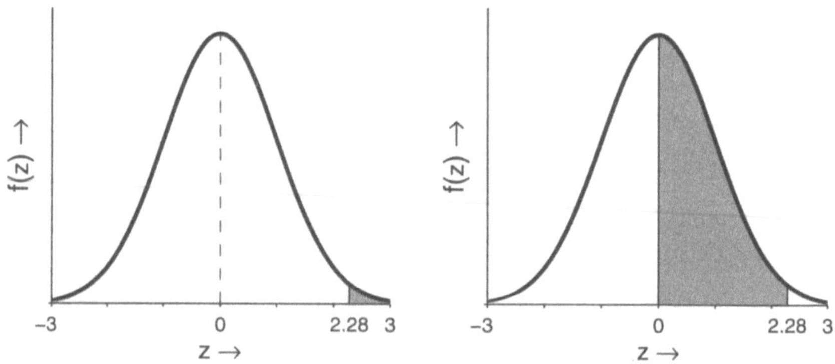

Figure 2.2. The z distribution, showing the probability areas, p, for $z \geq 2.28$ (left panel) and $0 \leq z \leq 2.28$ (right panel)

In fact we tested two hypotheses:

H_0 The mean recognition time of 'rente' does not differ from 490 ms; the difference found is simply due to sampling variability.

H_1 The mean recognition time of 'rente' differs from 490 ms.

The one-sample z test has the following components:

COMPONENTS OF ONE-SAMPLE Z TEST

- The statistic of interest: $\bar{X}$.
- The hypothesized effect in the population if H_0 is correct: $\mu = 490$.
- The standard error:

$$SE_{\bar{X}} = \frac{\sigma}{\sqrt{n}} \qquad (5)$$

  where n is the size of the sample, and σ the standard deviation in the population.
- σ is generally not available, it has to be estimated by s, the standard deviation of the sample. This estimation is correct if $n > 60$.

We found a small p value (0.011), which means that if H_0 is correct ($\mu = 490$ ms) we have a very small probability of obtaining a sample mean of 523 ms or more. Thus, we reject the null hypothesis in favour of the alternative one: H_1.

Beginners in statistics often ask what the difference between one-tailed and two-tailed tests is. Here goes: The first step is always to determine the probability of a type I error (= incorrectly rejecting H_0) that one is willing to accept. Let us assume that α is set at 5%. The next step is to decide what you expect:

(a) Observations (like a mean value of the sample) which are larger than the mean value under H_0.
(b) Observations which are smaller than the mean value under H_0.
(c) Not bothering about larger or smaller observations, both larger and smaller mean values may confirm the alternative hypothesis.

In cases (a) and (b) the odds are on one side and you use an α of 5% on this one side of the distribution (one-tailed testing). Under (c) you distribute α of 5% over the two tails of the distribution: 0.025 are the critical probability values now.

We have revised z values. This puts us into the position where we can use Standard Errors to assess the accuracy of our estimates of population parameters. Let us focus, once again, on estimates of μ on the basis of a sample with $\bar{X}$ and standard deviation s. It would be nice if we could qualify the accuracy of our estimate, for instance by saying: There is a 95% probability that μ is within an interval of y units around the sample mean $\bar{X}$.

Such a qualification of an estimate can be given on the basis of the following logic:

- We know the distribution of $\bar{X}$ if σ and n are known. It is a normal distribution with $\mu_{\bar{X}} = \mu$, which can be standardized.
- We can determine the probability to obtain values of $\bar{X}$ between $\bar{X} \pm C$ (C is an arbitrary constant). The resulting interval is called the confidence interval (= *CI*).
- By virtue of $z = (\bar{X} - \mu)/(\sigma/\sqrt{n})$ we can express the difference of $\bar{X}$ from μ by $z \times (\sigma/\sqrt{n})$.
- If we set the probability α to 5%, there is a $(100 - 95)\%$ probability that the $\bar{X}$ values will *not* miss μ by more than $z_{\alpha/2} \times (\sigma/\sqrt{n})$. α is distributed over the two tails; thus the z values are associated to $\alpha/2$.
- Conversely, if we have a sample mean $\bar{X}$, there is a probability of 95% that μ is included in the interval $\bar{X} \pm z_{\alpha/2} \times (\sigma/\sqrt{n})$.

Figure 2.3 depicts the sampling distribution of $\bar{X}$, on the basis of $\mu = 300$, $\sigma = 40$, and $n = 100$; the $SE = 40/10 = 4$.

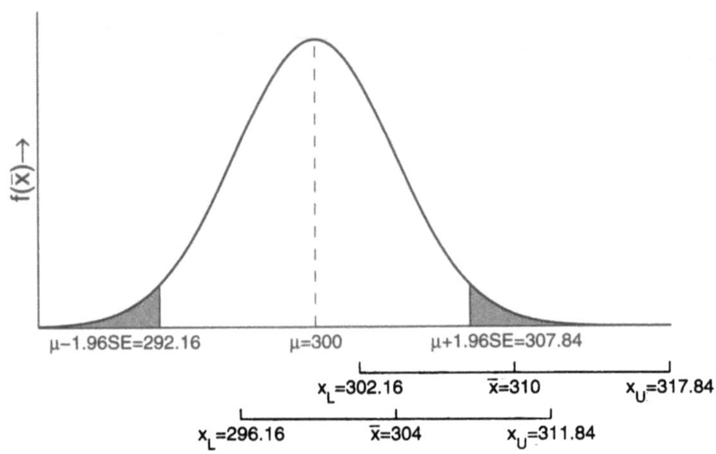

Figure 2.3. Sampling distribution of $\bar{X}$ on the basis of $\mu = 300$, $\sigma = 40$ and $n = 100$, together with the 95% confidence interval (= *CI*) around $\bar{X} = 304$ and 310

The interval within which $\bar{X}$ is located with a probability of 95% is (with z corresponding with 2.5% = −1.96 and + 1.96):

$$300 \pm 1.96 \times \frac{40}{\sqrt{100}} \rightarrow 292.16 \leq \bar{X} \leq 307.84$$

Let us now approach the problem from the opposite direction. If we observe an $\bar{X}$ just outside the critical values of the distribution (for instance $\bar{X} = 310$, just outside the 2.5% point 307.84), we see that the μ is not included in the 95% confidence interval, which is $\bar{X} \pm 1.96 \times \sigma/\sqrt{n}$ $(\sigma/\sqrt{n} = SE_{\bar{X}})$. Apparently '$\bar{X}$ being in the critical region of the distribution around μ' is equivalent to 'μ not being included in the confidence interval around $\bar{X}$'. Thus we can define the $(1-\alpha)\%$ confidence interval as:

$$\bar{X} - z_{\alpha/2}SE_{\bar{X}} \leq \mu \leq \bar{X} + z_{\alpha/2}SE_{\bar{X}} \qquad (6)$$

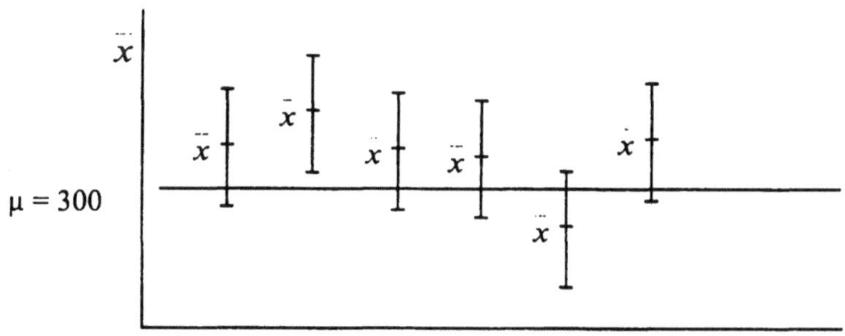

Figure 2.4. Sample means $\bar{X}$ and associated confidence intervals in relation to μ=300

In Figure 2.4, we show a number of sample means $\bar{X}$ and associated confidence intervals. The samples are taken from a population with $\mu = 300$. Most of the time μ is included in the confidence interval, but occassionally it is not (in 5% of the cases if it is a 95% confidence interval).

2.4. The t distribution

We have to mention a small but crucial problem here. If we do not have many observations in our sample (less than 60), and the standard deviation(s) in the population(s) have to be estimated – which is nearly always the case – the z values we calculate do not have an exact standard normal distribution, but a slightly different one, the t distribution, with larger 'tails'. This can be seen in Figure 2.5.

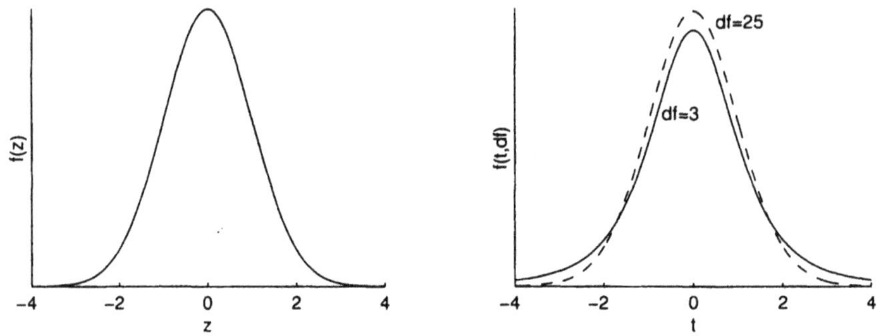

Figure 2.5. The z distribution (left panel) and two t distributions, with $df = 3$ and $df = 25$ (right panel)

Unlike the z distribution, the form of the t distribution varies as a function of the number of observations, more precisely, the distribution varies as a function of the number of *degrees of freedom*. Degrees of freedom (= df) is an important concept in statistics. In fact it is nothing more than the number of free choices you can make if the sum(s) (or mean(s)) of one or more series of observations are fixed. If you have the series 1, 7, 8, 5, 5 and the mean is 4, then four choices can be made freely: Having chosen 1, 7, 8 en 5 the last number must be 5 in order to keep the mean of 4. If you have two of these series of numbers, then the df are $2 \times (5-1) = 8$. Our first example in Section 2.3, however, was a marginal one; the df was $28 - 1 = 27$. When we compare the p value for $z = 2.28$ ($p = 0.011$) with the p value obtained with the t value, namely 0.015, we observe a slight difference. With increasing values of df, the t distribution becomes very close to the z distribution. That is why we use t values as our default as all statistical packages do. The formulas do not differ, only the p values associated with t or z may do, just by 'table look-up'. The difference between the ts with $df = 3$ and $df = 25$ (right panel), and the z distribution (left panel) is illustrated in Figure 2.5.

Below we reproduce the critical t values (at which $p(t \geq T)$ are 0.10, 0.05, 0.025, 0.010 and 0.005) for two values of degrees of freedom (60 and 120).

Table 2.3. Critical t values for df = 60 and 120

df	0.10	0.05	0.025	0.010	0.005
60	1.296	1.671	2.000	2.390	2.660
120	1.289	1.658	1.980	2.358	2.617

2.5. Statistical power

The aim of inferential statistics is to make statements about population(s) on the basis of restricted information, i.e. the information contained in sample(s) which are drawn from the (for obvious reasons much larger) population(s). The statements we make in the context of inferential statistics are always conditional. An example: If we say that we found a statistically significant result ($\alpha = 0.05$) on the basis of a t test, we mean that in less than 5% of the cases we would have found a value for t equal to or higher than the observed one if the mean of the population (μ) had had the hypothesized value (= the condition). The α level (in fact $1 - \alpha$) indicates the extent to which we have confidence in the assumption that the two population means do indeed differ. What about the situation in which we do not find a statistically significant result? Quite often researchers restrict themselves to reporting 'not significant at the 5% level'. However, in the case of non-significant results, we ought to add another piece of information: What chance have we given to the effect to be detected by our t test on the basis of the sample(s) at hand? In fact, this kind of information is similar to the information a biologist has to give if (s)he tries to detect an effect in a sample of fluid: What was the strength ('power') of the microscope, and under which conditions did (s)he obtain the sample?

In Table 2.4 we schematize the key concepts of this section. Four probabilities play a role:

- α = the probability of incorrectly rejecting H_0 (Type I error): $p(\text{rejecting } H_0 | H_0 = \text{true})$.
- β = the probablity of incorrectly accepting H_0 (Type II error): $p(\text{failing to reject } H_0 | H_0 = \text{false})$.
- $1 - \alpha$ = the probability of failing to reject H_0 when H_0 is true: $p(\text{failing to reject } H_0 | H_0 = \text{true})$.
- $1 - \beta$ = the probability of rejecting H_0 when it is false (power): $p(\text{rejecting } H_0 | H_0 = \text{false})$.

Table 2.4 illustrates that the probabilities discussed here are all *conditional* probabilities.

Table 2.4. The probabilities of the possible outcomes in hypothesis testing

		in the population	
		H_0 is true	H_0 is false
decision	Fail to reject H_0	$1 - \alpha$	β Type II error
	Reject H_0	α Type I error	$1 - \beta$ Power

Let us look at an example on the basis of a binomial variable. A binomial variable is a variable that has just two outcomes: 'yes' or 'no'; tossing a coin yields outcomes on the binomial variable 'head' or 'coin'. Likewise we might wonder what the probability is of accepting the hypothesis that a specific assimilation process takes place in 50% of the cases (equivalent to H_0: $\pi = 0.50$), whereas the real chance of that process happening is 70%; (H_1: $\pi = 0.70$): the *effect size* Δ is 20%. The process can be observed in 40 independent cases. Figure 2.6 illustrates the situation, with α set at 0.05.

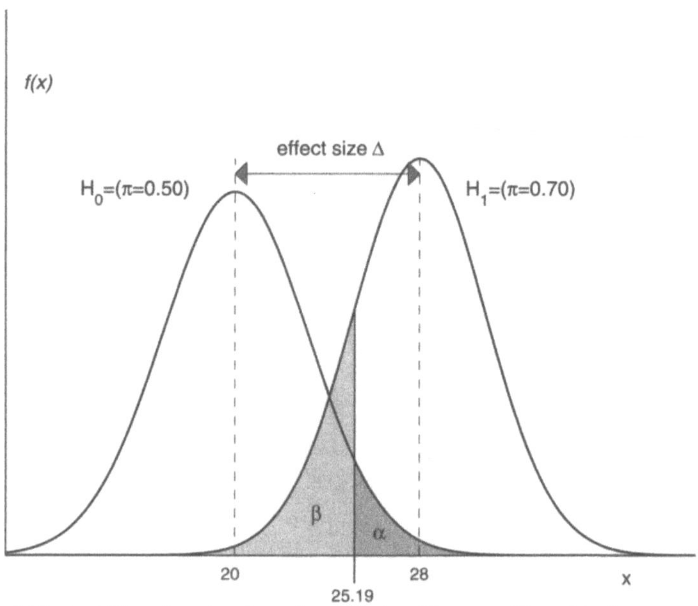

Figure 2.6. Probability density functions under H_0, $n = 40$, with the expected value $\pi = 0.50$, and the actual value of $\pi = 0.70$

It is important to remember that when both $n\pi$ and $n(1 - \pi)$ are greater than or equal to 5, we can use the normal distribution with $z = (X - \mu)/\sigma$.

For a binomial variable with $\mu = n\pi$ and $\sigma = \sqrt{n\pi(1-\pi)}$ the following calculations can be made (for clarity's sake, the continuity correction – which has to be applied when we switch from discrete to continuous variables – is omitted here; the correction consists in adding 0.5 to X if $X < n\pi$) and subtracting 0.5 if $X > n\pi$):

1. Under H_0: $\mu = 40 \times 0.50 = 20$, $\sigma = \sqrt{40 \times 0.50 \times (1-0.50)} = 3.162$. Thus the critical value of z corresponding to $\alpha = 0.05 = 1.64$; $1.64 = (X-20)/3.162$. This results in a critical value of X of 25.19.
2. Under H_1: $\mu = 40 \times 0.70 = 28$ (remember: here $\pi = 0.70$); $\sigma = \sqrt{40 \times 0.70 \times (1-0.70)} = 2.90$.

Figure 2.6 depicts the probability density functions associated with H_0: $\pi = 0.50$ and H_1: $\pi = 0.70$. As the probability surfaces under the curves are 1, but the standard deviations differ, the curves do not have equal heights. We can now determine the z value of the H_1 distribution which corresponds to $X = 25.19$ (the critical value under the H_0 distribution):

$$z = \frac{(25.19 - 28)}{2.90} = -0.969 \qquad (7)$$

This corresponds to a p value of 0.166. The corresponding power is $1 - 0.166 = 0.834$. This value stands for the probability that we detect an existing effect of 0.20 ($= 0.70 - 0.50$) in the population on the basis of a sample of 40 observations.

A technical remark can be added: We had to calculate z twice, once under H_0 and the second time under H_1; both μ and σ of binomial variables differ as a function of π, so does z. The only 'stable' value is the critical value on the X scale, which serves as an intermediate between z under H_0 and z under H_1.

Let us now assume that under H_1 π is 0.60 instead of 0.70. The effect size is here reduced to 10% (60% – 50%). The situation is illustrated in Figure 2.7.

Under H_0 the X value which corresponds to $z = 1.64$ is still 25.19. We now determine the z values which correspond to $X = 25.19$ under the H_1 distribution (n is still 40): $\mu = 40 \times 0.60 = 24$; $\sigma = \sqrt{40 \times 0.60 \times (1-0.60)} = 3.10$; $z = (25.19 - 24)/3.10 = 0.384$. The corresponding p value is 0.35 and the power is $1 - 0.35 = 0.65$, which is less than in the case of an alternative hypothesis where $\pi = 0.70$. We invite the reader to calculate the power in this example when n is reduced from 40 to 20. We have seen that the power of a

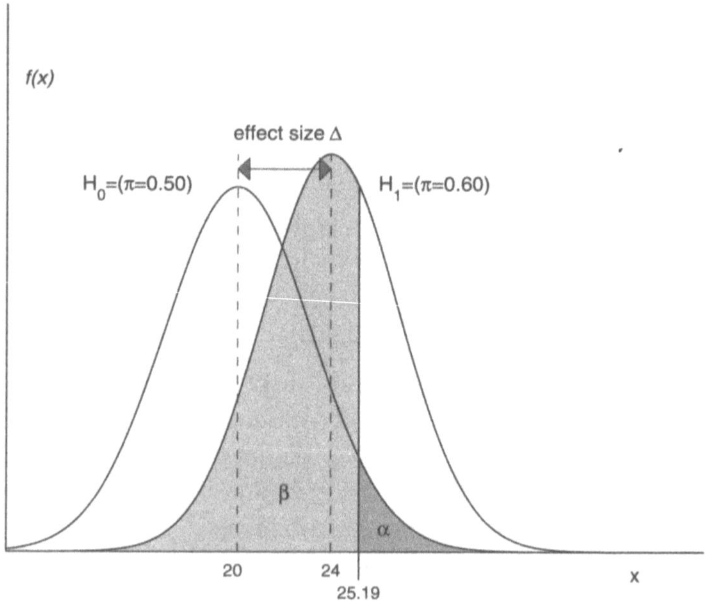

Figure 2.7. Probability density functions under H_0, $n = 40$, with the expected value $\pi = 0.50$, and the actual value of $\pi = 0.60$

test in any case depends on:

1. the effect size Δ: the difference to be detected between two population means
2. the adopted α level
3. the number of observations n.

2.6. Determining the sample size needed

To determine the achieved power on the basis of a given effect size Δ, n, and σ is one thing, but a more productive activity is to determine the sample size needed in order to achieve a specified power value. As an example, we determine the sample size needed to achieve a pre-determined power in a one-sample case with a binomial variable. The effect that has to be detected is $\pi_1 - \pi_0 = 0.70 - 0.50 = 0.20$; the significance level adopted is $\alpha = 0.05$ (one sided), and the power to be achieved is 0.80. To that purpose we can solve n from a set of two equations. The first equation has $z = 1.645$ on its left side, with $\alpha = 0.05$. The z value of the second equation corresponding to a probability surface of 0.20 (power = 0.80 = $1-\beta$) is -0.842.

$$
\begin{array}{llll}
H_0 : z = & 1.645 = (X-0.5n)/(0.5\sqrt{n}) & \longrightarrow & X = 0.5n + 0.8220\sqrt{n} \\
H_1 : z = & -0.842 = (X-0.7n)/(0.458\sqrt{n}) & \longrightarrow & X = 0.7n - 0.3856\sqrt{n} \\
\hline
 & & & 0 = -0.2n + 1.2076\sqrt{n} \\
 & & & 0.2n = 1.2076\sqrt{n} \\
 & & & (0.2n)^2 = 1.4583n \\
 & & & 0.04n = 1.4583 \\
 & & & n = 36.46
\end{array}
$$

In these calculations we obtained terms like $0.458\sqrt{n}$ on the basis of the following operation:

$$\boxed{\sigma = \sqrt{n\pi(1-\pi)} = \sqrt{n \times 0.70 \times 0.30} = \sqrt{0.21n} = 0.458\sqrt{n}}$$

Of course, in practice we *never* need to solve this kind of equation. Tables are available and also software (we mention SAMPLEPOWER provided by SPSS and NQUERY ADVISOR). SAMPLEPOWER nearly yields the same result (n = 37) as our calculations by hand, in which rounding has to be carried out to the next integer. The output of SAMPLEPOWER is given in Table 2.5.

Table 2.5. (Part of the) SPSS output of SAMPLEPOWER

One-sample proportion

	Proportion Positive	n of Cases	Standard Error	95% Lower	95% Upper
Population	0.70	37	0.08	0.57	0.81

In the preceding sections we have seen that the power of a test depends on a number of factors: The adopted α level, the effect size to be detected, the number of observations in the sample, and the standard deviations of the populations from which the samples are drawn, as they determine the Standard Errors of the sampling distributions involved. We give the following list of ways to increase the power of a test:

1. Increase α. The consequence is a greater probability of incorrectly rejecting a H_0, and a smaller probability of incorrectly accepting H_0.
2. Increase the difference (size of the effect) we would like to detect: Δ. Mere intuition tells us that a greater effect is more easily detected in the midst of error and noise than a small effect.

3. Reduce σ. This reduces the *SE*, on the basis of (example) $\sigma_{\bar{X}} = \sigma_X / \sqrt{n}$.
4. Increase n; see above.

For clarity's sake we used absolute measures for values of effect size: the difference between π_2 and π_1, the difference between two mean values $\bar{X}_2 - \bar{X}_1$ etc. However, absolute differences are not the best measures of effect sizes. Rather, when effect sizes are reported they should be expressed as ratios of absolute differences and measures of (common) standard deviations (Cohen 1988). As an example, the effect size of the difference between two means is:

$$d = \frac{\bar{X}_2 - \bar{X}_1}{s_{\text{common}}} \qquad (8)$$

Every statistic has its own definition for the effect size. Cohen provided guidelines to assess effect sizes:

value	*label of effect size*
0.20	small
0.50	medium
0.80	large

As an illustration we give in Table 2.6 the power achieved with different sample sizes to detect the difference between $\pi = 0.50$ and $\pi = 0.70$, and the difference between $\pi = 0.50$ and $\pi = 0.80$, one-tailed with α set at 0.05.

Table 2.6. Samples sizes and achieved power to detect the difference between (a) $\pi_0 = 0.50$ and $\pi_1 = 0.70$, and (b) $\pi_0 = 0.50$ and $\pi_1 = 0.80$

$\pi_0 = 0.50$, $\pi_1 = 0.70$		$\pi_0 = 0.50$, $\pi_1 = 0.80$	
sample size	achieved power	sample size	achieved power
$n = 10$	0.37	$n = 10$	0.65
$n = 20$	0.58	$n = 20$	0.89
$n = 30$	0.73	$n = 30$	0.97
$n = 40$	0.83	$n = 40$	0.99
$n = 50$	0.90	$n = 50$	1.00
$n = 60$	0.94	$n = 60$	1.00

2.7. Suggestions for statistical reporting

EXAMPLE 1

The mean duration of a sample of 100 measurements of the vowel [a] in identical prosodic positions is 120 ms. The standard deviation of the measurements is 35 ms, the boundaries of the 95% CI are 113.1 and 128.9 ms.

EXAMPLE 2
The designer of therapy X assumed that 70% of the patients would declare themselves cured after therapy. A reseacher assumed that the curing percentage was around 50%. Thus the effect to be detected was 20%; the α level (one-sided) was set at 5%, and the power of the test at 80%. The number of observations was accordingly set at 28.

2.8. Terms and concepts

- α *level*: The risk one is willing to accept the alternative hypothesis (H_1) while the null hypothesis (H_0) is true. Quite often a risk of 5% ($p = 0.05$) is accepted.
- β *level*: The risk one is willing to accept the null hypothesis while the alternative hypothesis is true. Although quite often a risk of 20% ($p = 0.20$) is accepted, the β value depends on the actual risks involved in incorrectly accepting the null hypothesis.
- *Degrees of freedom*: The number of observations that can be 'freely' varied after having determined the overall mean or sum of a sample.
- *Effect size*: The difference between the mean of the distribution of outcomes under the null hypothesis and the mean of the distribution of outcomes under the alternative hypothesis that are to be detected.
- *Power of a test*: 1-β.
- *Standard error*: Standard deviation of the distribution of sample statistics.

2.9. Exercises

1. Determine whether the following values of z should be regarded as significant at the 5% level, a) one-tailed, b) two-tailed: $z = 2.80$; $z = 1.65$; $z = -1.75$.
2. In which situations should one-tailed and two-tailed tests be carried out?
3. Do larger sample sizes decrease or increase the power of a test?
4. Does the increase of type II errors decrease or increase the power of a test?
5. It has been hypothesized that word items of a specific type are recognized

in 500 ms or less. An experiment was carried out with 35 participants who were presented with an item of that type. The mean Reaction Time was 505 ms; the standard deviation was 20 ms. Is there any reason to assume that the hypothesis was incorrect? Use a significance level of 1%.

6. A phonetician wants to test the hypothesis that in condition C /t/ is more often realized as [t] than as [d]. An effect size of 15% or more is regarded as a support for this hypothesis. The H_0 is that in condition C /t/ is realized equally often as [t] or [d]. A sample of 40 occurrences of /t/ is obtained. Calculate the power of the test when α (one-tailed) is set at 5%.
7. A phonetician wants to test the hypothesis that in condition C /t/ is realized as [t] rather than as [d]. An effect size of 10% is regarded as a support for the hypothesis. The power of the test should be 80%, a significance level of 5% is adopted (one-tailed). Calculate the size of the sample needed to achieve the desired power.
8. What are the p values associated with $z \geq 1.64$ and with $t_{29} \geq 1.699$?

Chapter 3
Basic statistical procedures: two samples

3.1. Preview

In many cases we are not interested in tests in which we compare the outcome obtained in one particular condition and the hypothesized outcome (the *one-sample test*), but rather in outcomes obtained in two (or even more) conditions. In this chapter we review basic statistical procedures used to compare mean values measured at interval scale level in two conditions. The mean values can be acquired from participants who are randomly assigned to different conditions (the *independent samples* design; Section 3.2)), but also from participants who are tested in two conditions (the *dependent samples* or *dependent samples* design; Section 3.3). It is absolutely crucial to use the tests appropriate to the two different designs. In Section 3.4 we show how to carry out t tests with SPSS. Section 3.5 discusses the analysis of differences between proportions, a frequently occurring situation in language research. We also devote sections to statistical power (Section 3.6) and sample sizes (Section 3.7), which are concepts of major concern in the design of an experiment or a survey. Why carry out an experiment with insufficient observations to detect a predetermined effect size with a specified probability?

3.2. Hypothesis testing with two samples

Let us assume that a particular word, for instance 'rente' is presented to participants in two different conditions: in condition 1 the word is presented auditorily, in condition 2 visually. The task of the participants is to press a button as soon as they detect the word. It was decided to assign the participants randomly to one of the two conditions. For some reason only 28 responses became available in condition 1 and 37 in condition 2. We are interested in the difference between the mean values of the Reaction Times (RTs) obtained in the two conditions. In condition 1 the mean RT was 523 ms ($s = 76.44$), in condition 2 it was 476 ms ($s = 70.10$). The question is whether there is a difference in the time taken to recognise 'rente' with visual and auditory presentations. With the hypothesis testing approach, we can set up the following hypotheses:

H_0 There is no difference in the mean recognition time of 'rente' in visual and auditory presentations.

H_1 There is a difference in the mean recognition time of 'rente' in visual and auditory presentations.

The two-sample z test is comparable to the single sample one. There are the following components:

COMPONENTS OF THE TWO-SAMPLE Z TEST:

– The statistic of interest: the difference $\bar{X}_2 - \bar{X}_1$.
– The hypothesized effect in the population if H_0 is correct: $\mu_2 - \mu_1 = 0$.
– The standard error:

$$SE_{\bar{X}_2 - \bar{X}_1} = \sqrt{\frac{\sigma_2^2}{n_2} + \frac{\sigma_1^2}{n_1}} \qquad (1)$$

where n_1 is the size of the first sample and n_2 the size of the second sample; σ_1^2 and σ_2^2 are estimated by s_1^2 and s_2^2 respectively.

Let us assume that H_0 is true: $(\mu_1 = \mu_2)$. We draw a large number of samples from these two distributions, and calculate for each sample the difference between the two sample means $(\bar{X}_2 - \bar{X}_1)$. If we make a graph of all the differences between sample means, we get a distribution with a mean value of zero, and *SE* as standard deviation, as shown in Figure 3.1.

Obviously the value of the *SE* depends both on the value of n and of the standard deviations of the populations from which the samples are drawn. The larger n is, the smaller *SE* becomes; the larger the values of s are, however, the larger *SE* will be. This is intuitively evident: When the sample sizes are large, the differences between the two sample means do not vary much, but if the populations from which the samples are drawn have large standard deviations, we can expect varying values of sample means, and consequently varying values of differences $\bar{X}_1 - \bar{X}_2$. The z statistic for the two-sample case becomes:

$$z = \frac{(\bar{X}_2 - \bar{X}_1) - (\mu_2 - \mu_1)}{\sqrt{\sigma_2^2/n_2 + \sigma_1^2/n_1}} = \frac{\bar{X}_2 - \bar{X}_1}{\sqrt{s_2^2/n_2 + s_1^2/n_1}} \qquad (2)$$

In our example, $n_1 = 28$, $n_2 = 37$. The values of σ are estimated by $s_1 = 76.44$ and $s_2 = 70.10$. Thus we obtain an $SE_{\bar{X}_2 - \bar{X}_1}$ of 18.479.

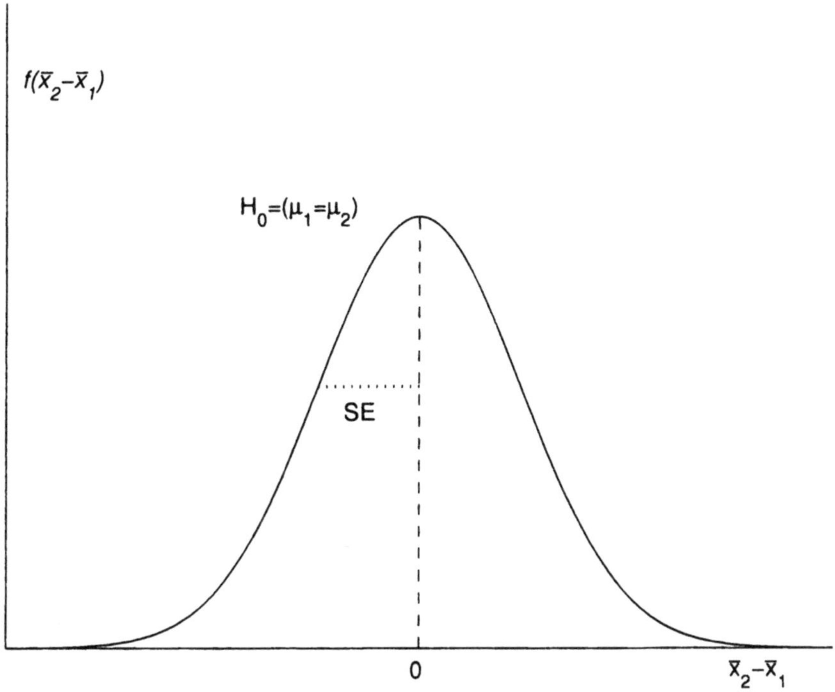

Figure 3.1. Distribution of $\bar{X}_2 - \bar{X}_1$ under H_0: $\mu_2 = \mu_1$

The two-sample z test statistic for this example is:

$$z = \frac{523 - 476}{\sqrt{76.44^2/28 + 70.10^2/37}} = 2.543 \qquad (3)$$

The associated p value is less than 0.01. Thus we found confirmation for the alternative hypothesis.

If the number of observations in one or both groups is small ($n_1 < 30$, and/or $n_2 < 30$), we have to use an 'independent samples t test' instead of a z test. The effects are, as was the case in the one-sample situation, a) a different 'table look-up', and b) the use of degrees of freedom. The formula for z is not affected. That is why one has not to specifiy in SPSS whether one wants a t test or a z test. By default a t test is applied. With larger numbers of observations the results of both tests are identical. With the t test for two independent samples, the concept of degrees of freedom may become clearer. In Chapter 2, with one-sample t tests, the df was $n - 1$. Why not use n therefore and shift the lines of the t table by 1? This would have been a good idea if we did not use t tables for other purposes too. The t test for two independent samples provides a good example. The df for that test $= n_1 - 1 + n_2 - 1 = n_1 + n_2 - 2$. Assume that $n_1 = 10$ and $n_2 = 7$. The total n is 17, but we have to use $df = 15$. If the total n in a one-sample test is also 17 the corresponding $df = 16$. Thus we see that just mentioning n as the line which directs table look-up for t, is not a good idea as it would direct us to the line labelled '17' in both cases, which is wrong. Using dfs avoids any possible confusion and directs us to the right value of the parameter which determines the shape of the t distribution: the dfs, which is $n - 1$ for the one-sample test and or $n_1 + n_2 - 2$ for the two-sample test.

In Table 3.1 we present two data sets A and B. In Table 3.2 the results of an analysis with the Compare Means procedure of SPSS on data set A are presented (for more details on SPSS see Section 3.4). Table 3.3 contains the outcomes for data set B.

Table 3.1. Two data sets with independent observations obtained in two groups each: 'gp1' and 'gp2'

	data set A		data set B	
	gp1	gp2	gp1	gp2
	15	7	12	10
	16	8	22	3
	18	10	12	11
	13	5	19	3
	10	9	7	12
$\bar{X}$	14.4	7.8	14.4	7.8
s	3.05	1.92	6.02	4.44

The two data sets in Table 3.1 have the same means for the two groups, but the standard deviations are different. The standard deviations in B are larger. Let us look at the consequences. We carried out t tests on both data sets A and B. The results are given in Tables 3.2 and 3.3.

The SPSS output in Table 3.2 contains the following components:

① Std. Deviation: The standard deviation of the data in each of the groups.

② Std. Error Mean: Estimated by $s/\sqrt{n}$.

③ Levene's Test for Equality of Variances: This test assesses whether the assumption of the t test, the equality of variances in both groups, is warranted. This test considers the ratio s_1^2/s_2^2 which follows the F distri-

Table 3.2. (Part of the) SPSS output from the *t* test procedure (Compare Means, Independent-Samples T Test) applied to the dependent variable 'depa' of data set A in Table 3.1

Group Statistics

	gp	N	Mean	① Std. Deviation	Std. Error Mean ②
depa	1.00	5	14.4000	3.04959	1.36382
	2.00	5	7.8000	1.92354	.86023

Independent Samples Test

	Levene's test for ③ Equality of Variances	
	F	Sig.
depa Equal variances assumed	1.052	.335

	t-test for Equality of Means ④				
	t	df	Sig. ⑥ (2-tailed)	Mean Difference	Std. Error ⑤ Difference
depa Equal variances assumed	4.093	8	.003	6.6000	1.61245
Equal variances not assumed	4.093	6.478	.005	6.6000	1.61245

bution (see Chapter 4). If it is *not* significant, the variances are assumed to be equal, which is the case for our example: $F = 1.052, p = 0.335$.

④ t-test for Equality of Means: As we can assume that the variances are equal, we use the corresponding line in the table: $t_8 = 4.093$, $p = 0.003$. If the variances had not been equal (and as a consequence Levene's test significant), we should have used the results listed under 'Equal variances not assumed', which boils down to an adjustment of the degrees of freedom (here from $df = 8$ to 6.748).

⑤ Std. Error Difference: $SE_{\bar{X}_1-\bar{X}_2} = \sqrt{(s_1^2/n_1) + (s_2^2/n_2)}$.
In our case $\sqrt{3.0496^2/5 + 1.9235^2/5} = 1.612$.

⑥ Sig. (2-tailed): If we do not specify the direction of our alternative hypothesis (for instance $\bar{X}_2 > \bar{X}_1$), the accepted error (α) is distributed over both tails. If for instance α is 0.05, in a two-tailed test the critical values of the z or t distributions corresponding with $0.05/2 = 0.025$ have to be used. The one-tailed significance is $0.003/2 = 0.0015$.

The results of the analysis for data set B, with the same means as obtained in data set A are different, though. The output is given in Table 3.3.

Table 3.3. (Part of the) SPSS output from the *t* test procedure (Compare Means, Independent-Samples T Test) applied to the dependent variable 'depb' of data set B in Table 3.1

Group Statistics

	gp	N	Mean	Std. Deviation	Std. Error Mean
depb	1.00	5	14.4000	6.02495	2.69444
	2.00	5	7.8000	4.4385	1.9849

Independent Samples Test

	Levene's test for Equality of Variances	
	F	Sig.
depb Equal variances assumed	.693	.429

	t-test for Equality of Means				
	① t	df	Sig. ② (2-tailed)	Mean Difference	Std. Error ③ Difference
depb Equal variances assumed	1.972	8	.084	6.6000	3.34664
Equal variances not assumed	1.972	7.354	.087	6.6000	3.34664

The SPSS output in Table 3.3 contains the following components showing the differences between data sets A and B:

① The result of the *t* test for data set B is no longer 4.093 as for dataset A, but 1.972.

② *t* is no longer significant.

③ This difference is due to the higher values of the standard deviations in the two groups of data set B, which raises the value of the *SE*.

The different results obtained with *t* tests for data sets A and B illustrate the effects of within-group variation on statistical testing.

3.3. Dependent samples and the t test

In the preceding example we considered two groups of data where different participants had looked at, or heard, the same word. If we had a situation where the *same* participants had looked at, or heard, the same words we could have used this extra information to allow us to improve our tests. Rather than considering the difference between the average reaction times for the two groups, we could look at the difference *within* each participant – how long did it take participant *a* to react to 'rente' with visual and auditory presentation? By examining the results for each participant, we can reduce the variability of our data – we are not interested in the differences between participant *a* and participant *b*, but rather in the difference between auditory and visual presentation. This allows us to focus on what we would like our experiment to investigate, and to bypass the variation in reaction time between participants.

The difference between *t* tests for paired and unpaired samples resides in the *SE* of the differences between $\bar{X}_1 - \bar{X}_2$ when the samples are paired (or matched or correlated) or unpaired. When they are unpaired the *SE* is:

$$SE_{\bar{X}_1-\bar{X}_2} = \sqrt{\frac{\sigma_1^2}{n_1} + \frac{\sigma_2^2}{n_2}} \qquad (4)$$

whereas with paired samples in which $n_1 = n_2 = n$ the *SE* is:

$$SE_{\bar{X}_1-\bar{X}_2} = \sqrt{\frac{\sigma_1^2}{n} + \frac{\sigma_2^2}{n} - \frac{2r_{12}\sigma_1\sigma_2}{n}} \qquad (5)$$

In the preceding equation r_{12} is the correlation between the variables X_1 and X_2. Therefore it is crucial to know beforehand whether our data are paired or not. Table 3.4 contains an example of a small data set, which is analyzed both as paired (dependent) and unpaired (independent) samples. The dataset consists of values obtained from 5 participants of an experiment in two conditions, 'c1' and 'c2'.

Table 3.4. Data set with repeated measures on five participants and in two conditions: 'c1' and 'c2'

participant	c1	c2
1	3	4
2	6	8
3	1	2
4	4	4
5	7	9

Table 3.5 shows the following results of the SPSS paired-samples test:

① $t = -3.207, df = 4, p$ (two-tailed) = 0.003, with $r_{12} = 0.974, s_1^2 = 5.698$ and $s_2^2 = 8.797$. The latter figures can be found in the SPSS output, but we confine Table 3.5 to the results which are directly related to the paired t test.

② We get a value of 0.374 for the standard error of $\bar{X}_1 - \bar{X}_2$, obtained by using formula (5), as follows: $\sqrt{5.698/5 + 8.797/5 - (2 \times 0.974 \times 2.387 \times 2.966)/5} = 0.374$. Thus the t value for paired samples is $-1.2000/0.374 = -3.207$.

Table 3.5. (Part of the) SPSS output from the paired samples t test procedure (Compare Means, Paired Sample T Test) applied to the data set from Table 3.4

	Paired differences				
			②	95% Confidence Interval of the Difference	
	Mean	Std. Deviation	Std. Error Mean	Lower	Upper
Pair 1 c1 - c2	-1.2000	.83666	.37417	-2.2389	-.1611

① t	df	Sig. (2-tailed)
-3.207	4	.003

If we had used the t test for independent samples, we would have got $t = -0.705, df = 8$, p (two-tailed) = 0.501. The *SE* of the means of unpaired differences is 1.703. This value is much higher than that of the means of differences of paired samples (0.374), as the substantial differences between the participants are absorbed in the *SE* without any further correction $-2r_{12}\sigma_1\sigma_2/n$.

3.4. t tests in SPSS

In order to carry out a *t* test in SPSS we have to present the data in a proper data matrix format. The relevant parts of the SPSS data matrix are reproduced in Table 3.6. The matrix consists of two variables ('gp' and 'depa') and ten cases (the participants). Properties or variables are presented columnwise; cases are presented rowwise.

Table 3.6. SPSS data matrix representing data set A from Table 3.1

gp	depa
1	15
1	16
1	18
1	13
1	10
2	7
2	8
2	10
2	5
2	9

The following steps in the Point-And-Click (PAC) window of SPSS lead to a *t* test of the data presented in Table 3.6:

window	Data View	
	click on	Analyze
	go to	Compare Means
	go to +	
	click on	Independent-Samples T Test
window	Independent-Samples T Test	
	click on	the variable 'depa'
	insert	under Test Variable(s)
	click on	the variable 'gp'
	insert	under Grouping Variable
	go to	Define groups
	fill in	1 for group 1, 2 for group 2
	click on	Continue
	click on	OK

The output will appear in an output file on the screen. SPSS can also run in syntax mode, where the commands have a text format. Window commands can be translated into syntax format by the option Paste. By clicking on Paste (instead of OK), our SPSS example delivers the following syntax which is automatically transferred to a syntax file:

```
T-TEST
  GROUPS=gp(1 2)
  /MISSING=ANALYSIS
  /VARIABLES=depa
  /CRITERIA=CIN(.95) .
```

The first component, GROUPS, specifies the dependent variable which is split up into the groups defined by the variable 'gp' and the two value levels of that variable. The significance level (α) is set at 0.05 (1 – 0.95) by the criteria specification. Syntax commands can be set to work by clicking on Run in the window of the syntax file.

The *t* test for dependent samples can be carried out as follows. The data matrix and the statistical procedure are different from the *t* test for independent samples. The data set in Table 3.7 contains five participants and a measure repeated in two conditions. As before, properties or variables are presented columnwise; cases are represented rowwise. This means that the two variables in Table 3.7 represent the two conditions. There are two variables (measures) for each participant, 'c1' and 'c2'.

Table 3.7. Relevant parts of SPSS data matrix representing the data from Table 3.4

c1	c2
3	4
6	1
1	2
4	4
7	9

The following steps lead to a *t* test of the data represented in Table 3.6 in SPSS:

window	Data View	
	click on	Analyze
	go to	Compare Means
	go to +	
	click on	Paired-Samples T Test
window	Paired-Samples T Test	
	click on	the variable c1
	click on	the variable c2
	insert	under Paired Variable(s)
	click on	OK

By clicking on Paste instead of OK our example delivers the following syntax which is automatically transferred to a syntax file:

```
T-TEST
  PAIRS= c1 WITH c2 (PAIRED)
  /CRITERIA=CIN(.95)
  /MISSING=ANALYSIS.
```

The first component, PAIRS, defines the two variables tested for their difference. The significance level (α) is set at 0.05 by the criteria specification. Syntax commands can be set to work by clicking on Run in the window of the syntax file.

3.5. Comparing two proportions

Many linguistic data come in a binary form: Was a particular word form used, or not; Did a speaker use a standard or non-standard allophone, etc. This type of data can be represented as a proportion, one count divided by the total number of occurrences. For example, in Standard Scottish English, one speaker realizes /x/ by [x] 40 times in conversation, and by [k] 20 times. The proportion of [x] is thus $(40/(40+20)) = 0.6667$ which can also be written as 66.67%. The proportion of usage of [k] is then $20/(20+40) = 0.3333$. We would like to know whether the proportion of usage of /x/ as [x] is higher than that of [k]. In other words, our alternative hypothesis is that p(realize [x]) > p(realize [k]), and the null hypothesis is p(realize [x])= p(realize [k]), which is the same as p(realize [x]) = 0.50. Again we can use a z score. We compare the observed proportion of occurrences of realized [x]s to the expected proportion (on the basis of the null hypothesis). The observed proportion is 0.6667, the expected proportion = 0.50. Again the standard error takes a specific form. This time it is:

$$SE_p = \sqrt{\frac{p(1-p)}{n}} \qquad (6)$$

where p is the sample proportion and n is the sample size. In this case, we have an estimated standard error of:

$$\sqrt{\frac{0.6667 \times 0.3333}{60}} = 0.0609$$

Thus we get:

$$z = \frac{0.6667 - 0.50}{0.0609} = 2.737 \qquad (7)$$

With a one-tailed test we assume that /x/ is more often realized as [x] than as [k], $p < 0.05$.

With two proportions, say another speaker of Standard Scottish English, the procedure is comparable to that used when comparing two means. Our second speaker was for instance recorded as realizing /x/ by [x] 20 times out of 50 uses (= 40%), and we are interested in the difference between the two rates. Our question now is: Does speaker 1 use [x] with a rate which is different from speaker 2? Our hypotheses are:

H_0 There is no difference in the rates of use of [x] between the two speakers.

H_1 There is a difference in the rates of use of [x] between the two speakers.

The standard error of the difference between these proportions is

$$SE_{p_1-p_2} = \sqrt{\frac{p_1(1-p_1)}{n_1} + \frac{p_2(1-p_2)}{n_2}} \qquad (8)$$

where p_1 is the proportion observed in the first sample, p_2 that observed in the second sample; n_1 is the first sample size and n_2 is the second. In our example the *SE* becomes:

$$SE_{p_1-p_2} = \sqrt{\frac{0.6667(1-0.6667)}{60} + \frac{0.40(1-0.40)}{50}} = 0.0922$$

The test statistic is:

$$z = \frac{p_1 - p_2}{SE_{p_1-p_2}} \qquad (9)$$

In this example, $0.2667/0.0922 = 2.8926$. We compare this value to percentage points of the normal distribution. The 95% points are −1.96 and +1.96, which are less extreme than our test statistic. Thus we get a p value smaller than 0.05 and we have evidence to reject the null hypothesis that there is no difference between the speakers.

3.6. Statistical power

Our considerations on 'power analysis' started in Chapter 2 with an example of a binomial variable. The advantage is that σ (the standard deviation) of the distributions of these variables is directly and solely dependent on π ($\sigma = \sqrt{\pi \times (1-\pi)}$) and n. Of course, this is not the case with other variables like, for instance, the interval variable milliseconds to measure durations of speech segments in phonetics, or reaction times in psycholinguistic research (remember that the standard deviation of the distribution of sample means is $\sigma/\sqrt{n}$). In case we are interested in the hypothesized difference between the means in two conditions, for instance the durations of [a] and [i], we may wonder what the power of our t test is if we measure 50 tokens of [i] and 50 tokens of [a]. Let us assume H_0 is true: ($\mu_1 = \mu_2$). We draw a large number of samples from these two distributions, and for each pair of samples we calculate the difference between the two sample means ($\bar{X}_2 - \bar{X}_1$). If we make a graph of all the differences between sample means, we get a distribution with a mean value of zero.

In Figure 3.2 we give the distributions of $\bar{X}_2 - \bar{X}_1$ under H_0 ($\mu_2 = \mu_1 = 80$) and H_1 ($\mu_2 = 95, \mu_1 = 80$), assuming $\sigma_1 = \sigma_2 = 30$. As in the binomial distributions (one distribution for each value of π, cf. Chapter 2) we see that the two distributions of $\bar{X}_2 - \bar{X}_1$ overlap. They create a set of values which are going to be interpreted as evidence for the H_0 whereas H_1 is true. The probability of this incorrect interpretation is β, which is 0.20. The power is $1 - 0.20 = 0.80$.

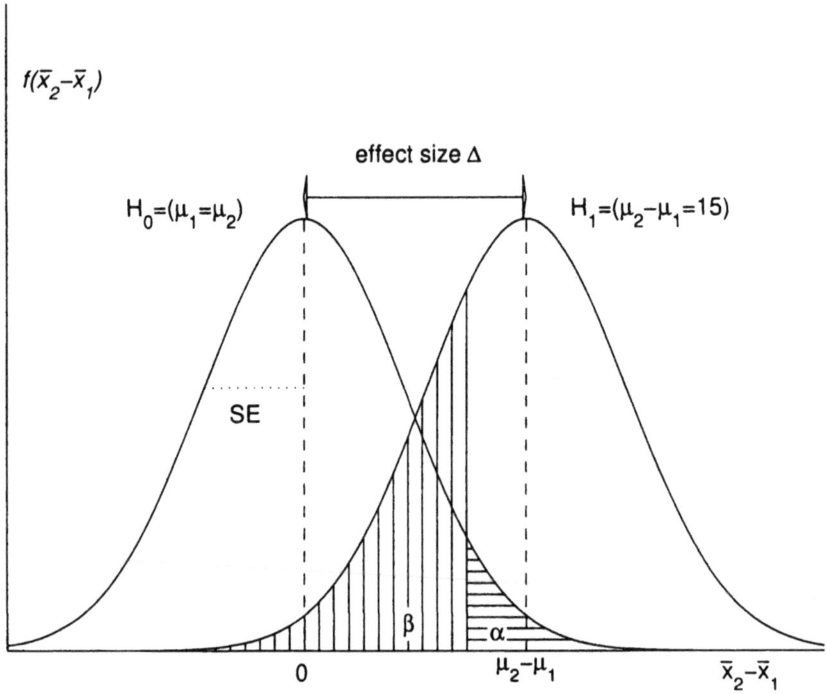

Figure 3.2. Distributions of $\bar{X}_2 - \bar{X}_1$ given $n = 50$ under the two hypotheses H_0 ($\mu_2 = \mu_1 = 80$) and H_1 ($\mu_2 = 95, \mu_1 = 80$), $\sigma_1 = \sigma_2 = 30$

In all basic textbooks on statistics the advice is given to calculate the power of the test to detect a predefined effect size, before carrying out the experiment or survey. We wonder whether anybody follows this advice, apart from researchers in medical settings. To carry out an experiment with a small power in a medical context is, in fact, a waste of energy and, quite often, of comfort of patients. That is why power analysis is still given more focus in the medical disciplines than in linguistic research. We think, however, that power analysis merits more attention in linguistic research, too.

3.7. How to determine the sample size

An obvious strategy, and this is the strategy followed in the medical sciences, is to determine the sample size needed to detect an effect size with a predetermined power. For this purpose we need:

1. the adopted significance level
2. the power which has to be achieved
3. the effect size to be detected, if present in the population
4. the standard deviations of the population(s) from which the sample(s) is (are) drawn (not necessary if we deal with a binomial variable: $\sigma = \sqrt{n\pi(1-\pi)}$).

We amend our example, but keep α at 0.05 (one-tailed), the required power at 0.80, and the $\sigma_1 = \sigma_2 = 30$. The effect size we would like to detect is $90 - 80 = 10$ ms. We want to know the sample size needed. The SAMPLEPOWER program yields 112 observations per groups, as can be seen in the output reproduced in Table 3.8.

Table 3.8. Output SAMPLEPOWER for $\alpha = 0.05$ (one-tailed), power = 0.80, $\sigma_1 = \sigma_2 = 30$, effect size = 10 ms, *t* test for independent samples

	Population Mean	Standard Deviation	N Per Group	Standard Error	95% Lower	95% Upper
Population 1	90.0	30.0	112			
Population 2	80.0	30.0	112			
Mean Difference	10.0	30.0	224	4.01	3.39	16.61
	Alpha = 0.0500, Tails = 1, Power = 0.8001					

There are a number of ways to increase the power of a test. This can be achieved by manipulating one or more of the factors which determine its magnitude: α (type I error), the effect size we would like to detect, the variation in the (sub)populations under discussion, n, the sample size(s). The latter two factors determine the standard error. We show the impact of these factors in Figure 3.3.

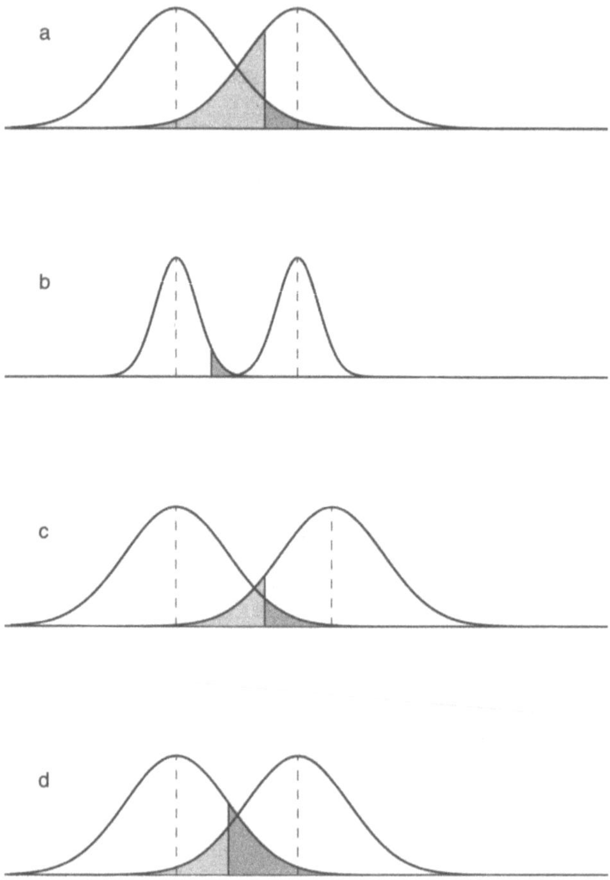

Figure 3.3. Distributions showing how different factors have an inpact on test power, a) default, b) smaller *SE*, c) larger effect size, d) larger value of α

Figure 3.3a shows a default situation, with two distributions, one associated with H_0, and one associated with H_1; α is set at 5%. In Figure 3.3b we show the effect of decreasing the *SE*, either by increasing n and/or a smaller σ. Figure 3.3c shows the effect of increasing the magnitude of the effect we would like to detect. Finally, Figure 3.3d demonstrates the effect of increasing α.

3.8. Suggestions for statistical reporting

EXAMPLE 1 (M = mean, SD = standard deviation)
For sample A we obtained M = 90.25 ms (SD = 12.16) and for sample B M = 105.50 ms (SD = 11.25). The difference was 15.25 ms. A t test for independent samples carried out on the data was significant at the 5% level (one-tailed, equal variances assumed): $t_{38} = 4.136$, $p < 0.01$.

EXAMPLE 2
The mean of sample A was 89.65 ms (SD = 11.33), of sample B 93.10 ms (SD = 5.96). The difference was 3.45 ms. A t test for independent samples carried out on the data was not significant at the 5% level (one-tailed, equal variances assumed): $t_{38} = 0.948$, $p = 0.118$. The observed power was 0.217.

3.9. Terms and concepts

- *Paired samples*: Observations on the same participant (subject) made in different conditions.
- *Power of a test*: $1 - \beta$.
- *Standard error of differences*: The standard deviation of the distribution of differences between two sample means $\bar{X}_1 - \bar{X}_2$.
- *Standard error of paired samples*: Differs from the standard error of the differences between two independent samples by the term $-2r_{12}\sigma_1\sigma_2/n$.

3.10. Exercises

1. Determine whether the following values of t should be regarded as significant at the 5% level, a) one-tailed, b) two-tailed: $t_4 = 2.80$; $t_{50} = 1.65$; $t_{14} = -1.75$
2. The following table gives the scores obtained in two conditions, 'c1' and 'c2':

c1	c2
1	3
2	5
1	1
2	5
7	14

Carry out two analyses with t tests (SPSS): a) imagine that the data are independent samples, b) imagine that the data are paired samples. Compare and discuss the results; use a one-tailed test with α set at 5%.

3. A researcher investigated the realization of /t/ in two conditions, 'c1' and 'c2'. In 'c1' it was realized as voiced in 7 cases out of 16, and in 'c2' in 22 cases out of 30. Does this difference support the hypothesis that there is a tendency that /t/ is realized voiced in 'c2' more often than in 'c1'? Use the 5% level, and do not use the continuity correction.
4. A clinical linguist wants to compare 'speech intelligibility' of two groups of young children with a cleft palate, Group 1 did not get a presurgical orthopaedic treatment, group 2 did. Intelligibility is assessed on the basis of a 7-point scale. The effect size is 1 scale point, with α set at 5% (one-tailed) and β at 20%. An estimate of σ in both subpopulations is 1.2 scale point. Determine the number of children in both groups who are needed to detect the effect size mentioned by using a program like SAMPLEPOWER or NQUERY ADVISOR.
5. An applied linguist wants to assess the effects of two different reading methods A and B. Method A is used in class 3A, method B in class 3B. The two classes involved are 'parallel' classes, for children with similar ages (approx. six years of age). In order to find out whether the ages differ between the two classes, a t test is carried out on the mean age differences (class 3A 6 years, 4 months old; class 3B 6 years, 6 months old). Why is it pointless to carry out a t test?

Chapter 4
Principles of analysis of variance

4.1. Preview

Analysis of variance is a statistical technique which is often applied in the analysis of language data. Its frequent use does not necessarily imply that it is a simple technique. Nevertheless, the basic concepts are quite straightforward. Analysis of variance can be applied if two or more groups of participants are compared on the basis of a so-called *dependent variable*. This may be the reaction time in a word recognition experiment, the performance in a language proficiency test or the acceptability scores of different types of computer speech. Two characteristics are essential. First, a participant belongs to one of the defined groups. Second, the performance of the participants on the dependent variable is measured by a score at the interval level. These subjects are introduced in Sections 4.2 and 4.3 on the basis of an important concept: the *model equation*. In Section 4.4 the *F distribution* is introduced, a key concept in analysis of variance and many other techniques. However, the application of analysis of variance often becomes more complicated. Complicating factors are, for instance, the introduction of more than one factor along which groups of participants are defined, the occurrence of repeated measures on the same participants (discussed in Chapter 8), the introduction of 'random' factors instead of 'fixed' ones, the use of 'nested' factors, and 'missing data' (see Section 4.4). In Section 4.5 we show how *one-way analysis of variance* is carried out with SPSS, i.e. an analysis with one independent variable. Section 4.6 is devoted to a practical issue: We would like to know what to do next when we obtain a significant outcome for an independent variable with more than two levels. In Sections 4.7 and 4.8 we discuss *power* and *sample size* in analysis of variance.

This chapter deals with the basic principles of analysis of variance, illustrated on the basis of straightforward examples. We keep complications for subsequent chapters. This chapter introduces the basic concepts of analysis of variance, starting with one-way analysis of variance, in which participants belong to a single set of different groups. The application is illustrated by examples from the statistical package SPSS.

4.2. A simple example

The basic ideas in analysis of variance can be illustrated by a simple example. Let us assume that we are interested in the effects of three methods of vocabulary learning in a second language. Suppose that three participants who have no knowledge of the second language in question, are randomly assigned to each of the methods ($3 \times 3 = 9$ participants). After the training period a vocabulary test is carried out. Table 4.1 contains two hypothetical data sets with test scores.

Table 4.1. Two hypothetical data sets of test scores, both for three groups of three participants

	data set A				data set B		
	group				group		
	I	II	III		I	II	III
	9	10	1		3	7	1
	1	2	5		4	6	2
	2	6	0		5	5	3
$\bar{X}$	4	6	2	$\bar{X}$	4	6	2

Both data sets in Table 4.1 have the same mean scores for the three groups. Given the outcomes in both data sets it is tempting to conclude that the participants belonging to group II performed better and, consequently, that the method used for that group is the method which should be preferred in vocabulary learning. On the basis of the outcomes of the first data set, however, such a conclusion is hardly warranted. Especially the large differences or variation in scores between the participants within the groups compared to the differences or variation in the mean scores between the groups cast serious doubts on the value of the differences between the mean scores. It is not unlikely that the variation in mean scores is brought about by the variation between the individual participants. In this case, the variation between the mean scores only reflects individual differences; it does not reflect differences between the three methods used. The second data set offers a sounder basis for the conclusion that the methods used produce different results, because the variation within the groups is limited. Because of the restricted amount of variation within the groups in comparison with the variation between the groups, it is more acceptable to reject the hypothesis that no differences occur between the groups. This line of reasoning takes us to the essence of analysis of variance: It involves the comparison of sources of variation. In our example two possible sources of variation can be distinguished, Variation brought about by effects induced by the so-called independent variable (the method), and variation which is not influenced by the independent variable and which is brought about by individual differences between participants: the error. In the following section (4.3) the elements which are necessary to draw a meaningful comparison between the two types of variation mentioned here are presented. Obviously it is not enough to just compare the magnitudes of variation measures (whatever these may be). In statistics we are not satisfied with statements like "variation A exceeds variation B". Instead, we look for a statistic that enables us to say with a specified (low) degree of uncertainty that the variation between group means is not based on chance, but on a genuine difference between two or more groups.

However, before giving more details of analysis of variance, we first have to answer an obvious question: Why should analysis of variance be used when it is also possible to carry out a series of t tests to detect differences between groups? The difference in mean scores of two groups can be statistically evaluated by means of a t test. If the t value calculated passes a specified probability level (often 0.05), the so-called null hypothesis, which states that there is no difference between the two mean scores observed, is rejected and the so-called alternative hypothesis that a real difference exists is accepted instead. Indeed, a t test can be applied to all pairs of groups if more than two groups are examined. If at least one of the t tests proves to be significant, the null hypothesis can be rejected. However sound this strategy may seem at first sight, something can go wrong with the so-called type I error. In applying a statistical test, a t test for instance, we run the risk of rejecting the null hypothesis while the null hypothesis is in fact true. The probability of a type I error is equal to the significance or α level (for instance 0.05). What happens to the α level if the test in question is repeated? The α level is not going to be 0.05, but much higher. This effect can be illustrated with an example based on elementary principles of probability theory.

Let us suppose that the probability of the first letter of a word being an 'e' is 0.10. What is the probability that at least one out of five words randomly selected from a text begins with the letter 'e'? It is:

$$1 - p(\text{not 'e'})^5 = 1 - (1 - .10)^5 = 1 - 0.90^5 = 0.41$$

An analogous mechanism is present in the repeated use of a t test on the same data set. Suppose we have gathered the scores of three groups of individuals and that t tests are applied at the 0.05 level (= α). The 0.05 level or 5% level is the probability of incorrectly rejecting the null hypothesis. If there are three groups we have to apply the t test three times. The probability of incorrectly rejecting at least one null hypothesis in k t tests is:

$$1 - p(\text{not rejecting the null hypothesis})^k = 1 - (1 - \alpha)^k \qquad (1)$$

If α is 0.05 and $k = 3$, that probability is $1 - (0.95)^3 = 1 - 0.86 = 0.14$. This value is well above the originally planned level of 0.05. In this last example the assumption is that the t tests applied use independent information. This is clearly not the case if the same samples are used in different tests; t tests are carried out on the data of samples I and II, I and III, and II and III. This implies that the real α level cannot be calculated precisely and that is the reason why a different approach should be chosen, i.e. analysis of variance.

4.3. One-way analysis of variance

In analysis of variance several terms are used to indicate the independent variable (representing the different groups compared) and its values. When we say that *treatment* j (representing one of the values of the independent variable) influences the magnitude of the scores, we sy that it constitutes an *effect*. If that treatment is one of a series of treatments, the series constitutes a *factor*, and the treatments are called the *levels* of that factor. Factors or independent variables are generally referred to by Greek letters: α, β etc.; the levels of the factors are indicated by subscripts. The first level or value of factor α is referred to as α_1; α_2 represents the second level of that factor and so on. The symbols are used to formulate models for the data or scores observed. In fact, formulating and testing models for the observed scores constitute the basic principle of analysis of variance, although the reports in which this technique is used often suggest otherwise. What is a plausible model of the data in Table 4.2? The scores can be symbolized by X_{ij}, in which i refers to the score of individual i and j to the level or treatment of the factor or independent variable investigated. So every score has a unique identity.

We can imagine a model in which every score X_{ij} equals an overall mean called μ. In such a model all scores observed are equal and the variation found is supposed to be caused by irrelevant individual differences between the participants. Clearly, the model $X_{ij} = \mu$ does not suffice for the data of Table 4.2 as the three subpopulation means differ from each other relatively

Table 4.2. Hypothetical data set for three groups of participants or treatments

	data set C		
	group I	group II	group III
	$X_{11} = 4$	$X_{12} = 6$	$X_{13} = 2$
	$X_{21} = 4$	$X_{22} = 6$	$X_{23} = 2$
	$X_{31} = 4$	$X_{32} = 6$	$X_{33} = 2$
$\bar{X}_j$	4	6	2

systematically. A term has to be included which represents the value added to or subtracted from the overall average, according to the specific treatment: $\alpha_j = \mu_j - \mu$. This yields the following extended model equation:

$$X_{ij} = \mu + \alpha_j \qquad (2)$$

Given the scores in Table 4.2, we can assign the αs the values 0, +2 and –2, a perfect match results between the scores predicted by the model and the scores observed. In real life, however, scores exhibit variation within a group of participants. The model needs to be extended further by a supplementary term representing an error component: ε_{ij}. It indicates the amount of random variation for participant i in group (or treatment) j. The full model equation of the scores takes on the following form:

$$X_{ij} = \mu + \alpha_j + \varepsilon_{ij} \qquad (3)$$

The error can take any value, whereas the treatment effects are assumed to have a constant value in each group. This last model can be applied to the two data sets of Table 4.1 using $\alpha_1 = 0$, $\alpha_2 = +2$ and $\alpha_3 = -2$ as estimated values. These values are obtained by $\bar{X}_j - \bar{X}$. The results are given in Table 4.3.

Table 4.3 shows how all individual scores are partitioned. A score is considered to be the sum of separate components. We assume specific values for μ and α_j. In empirical research, however, we do not know these values. They can only be estimated on the basis of the patterns present in the data. If the research design contains one factor or one series of treatments only, as in the examples given, the crucial question is if an effect α_j exists. It has to be decided which equation underlies the observed data, model 1 or model 2:

$$X_{ij} = \mu + \varepsilon_{ij} \qquad \text{model 1} \quad (4)$$

$$X_{ij} = \mu + \alpha_j + \varepsilon_{ij} \qquad \text{model 2} \quad (5)$$

Table 4.3. The full model (= model 2) specified for two hypothetical data sets of test scores (see Table 4.1), assuming that $\alpha_1 = 0$, $\alpha_2 = +2$ and $\alpha_3 = -2$

	data set A			data set B		
	I	II	III	I	II	III
	4+0+5 = 9	4+2+4 = 10	4−2−1 = 1	4+0−1 = 3	4+2+1 = 7	4−2−1 = 1
	4+0−3 = 1	4+2−4 = 2	4−2+3 = 5	4+0+0 = 4	4+2+0 = 6	4−2+0 = 2
	4+0−2 = 2	4+2−0 = 6	4−2−2 = 0	4+0+1 = 5	4+2−1 = 5	4−2+1 = 3
$\bar{X}_j$	4	6	2	4	6	2

What can be observed in the data is (1) the *variation* between the participants or informants *within* the groups and (2) the *variation* in mean scores or values *between* the groups. If there is no effect α_j, the variation observed between the mean scores of the groups (the between group variation) is brought about by the variation in the error component (the within group variation). If an effect α_j is present, the variation between the groups increases because of the differences between the groups. The variation between the groups now consists of error variation plus variation caused by the effect α_j. The question is how the variation between the groups can be compared with the variation within the groups. The customary measure for variation is the so-called variance, which in the case of a random sample of scores is defined as follows:

$$s_x^2 = \frac{\sum_{i=1}^{n}(X-\bar{X})^2}{n-1} = \frac{\sum_{i=1}^{n}x^2}{n-1} \qquad (6)$$

$\sum x^2$ is the sum of the squared deviations of X from the mean ($x = X - \bar{X}$) and n is the sample size. Normally, the term *Sum of Squares* (= *SS*) is used to refer to the deviation sum of squares. The sum of squares for the within group variation in the first group of data set A (see Table 4.1) is $\sum_{i=1}^{n_1}(X_{i1} - \bar{X}_1)^2 = (9-4)^2 + (1-4)^2 + (2-4)^2 = 38$; the *SS* for the second group is 32, and 14 for the third group. The *SS* between the groups is $\sum_{j=1}^{3}(\bar{X}_j - \bar{X})^2 = (4-4)^2 + (6-4)^2 + (2-4)^2 = 8$. In this example it is obvious that the sum of squares for the between group variation is smaller than the sum of squares for the within group variation. But how can both sums of squares be compared precisely, and how should the number of observations on which the sums of squares are based be taken into account? We have to find a statistic which applies to our data and whose distribution is known if the null hypothesis is true (= model 1: There is no effect α_j and the between group variation is equal to the within group variation). In order to arrive at this statistic, we have to decompose the observed (deviation) scores into parts. This can be done in the following way:

$$X_{ij} - \bar{X} = (\bar{X}_j - \bar{X}) + (X_{ij} - \bar{X}_j) \quad (7)$$

$\bar{X}$ is the general or 'grand' mean of all the observations in the different groups. The above equation is in fact a simple identity, but the interesting point is that the deviation score (on the left) is partitioned into a part which contains the difference between the group mean and the grand mean ($= \bar{X}_j - \bar{X}$), plus one which contains the difference between the individual score and its group mean ($= X_{ij} - \bar{X}_j$). The first part is the between group component, the second is the within group component. Squaring and summing this identity over n_j informants in the jth group results in the following:

$$\sum_{i=1}^{n_j}(X_{ij} - \bar{X})^2 = \sum_{i=1}^{n_j}(\bar{X}_j - \bar{X})^2 + \sum_{i=1}^{n_j}(X_{ij} - \bar{X}_j)^2 - 2\sum_{i=1}^{n_j}(\bar{X}_j - \bar{X})(X_{ij} - \bar{X}_j) \quad (8)$$

The right-hand part of the equation is obtained by using simple algebra: $(a-b)^2 = a^2 + b^2 - 2ab$, where a refers to the first term in the right-hand part of equation (8), and b to the second term. The last term on the right of formula (9) equals zero, as can be demonstrated in the following way: $\sum(\bar{X}_j - \bar{X})(X_{ij} - \bar{X}_j) = (\bar{X}_j - \bar{X})\sum(X_{ij} - \bar{X}_j)$; by definition $\sum(X_{ij} - \bar{X}_j) = 0$ and, consequently, the whole last term is 0. The first term of the right-hand part can be rewritten as $n_j(\bar{X}_j - \bar{X})$, the whole equation as:

$$\sum_{i=1}^{n_j}(X_{ij} - \bar{X})^2 = n_j(\bar{X}_j - \bar{X})^2 + \sum_{i=1}^{n_j}(X_{ij} - \bar{X}_j)^2 \quad (9)$$

By summing over k groups we obtain:

$$\begin{aligned} \sum_{j=1}^{k}\sum_{i=1}^{n_j}(X_{ij} - \bar{X})^2 &= \sum_{j=1}^{k} n_j(\bar{X}_j - \bar{X})^2 + \sum_{j=1}^{k}\sum_{i=1}^{n_j}(X_{ij} - \bar{X}_j)^2 \quad (10) \\ SS_{\text{total}} &= SS_{\text{between}} + SS_{\text{within}} \end{aligned}$$

The term in the left-hand part of the equation is the total sum of squares: The sum of squares of all the observations with respect to the grand mean $\bar{X}$. The right-hand part of the equation contains two terms: The first term represents the sum of squares between groups, the second one the sum of squares within groups. The *total sum of squares* can be partitioned into two additive and independent components: the *between groups sum of squares* (SS_{between}) and the *within groups sum of squares* (SS_{within}). Each sum of squares has to be divided by its associated degrees of freedom to obtain a variance estimate. The result is a so-called *mean square* (*MS*). The number of degrees of freedom for the total sum of squares is the total number of observations minus 1: $n-1$. One degree is lost because of the presence of the grand mean. The number of degrees of freedom for the within groups sum of squares is the number of observations within a group minus 1. Given k groups the number of degrees of freedom is $n-k$; k degrees are lost because of the presence of k group means. The number of degrees of freedom associated with the between groups sum of squares is the number of groups minus 1: $k-1$. One degree is lost here because of the presence of the grand mean. The degrees of freedom are additive too: $n-1=(k-1)+(n-k)$. The total number of degrees of freedom is equal to the number of degrees of freedom associated with the between groups sum of squares plus the number of degrees of freedom associated with the within groups sum of squares. The variance estimates for the between and within groups parts are obtained by dividing the respective sum of squares by their degrees of freedom:

$$s_b^2 = MS_{\text{between}} = \frac{\sum_{j=1}^{k} n_j(\bar{X}_j - \bar{X})^2}{k-1} \quad (11)$$

$$s_w^2 = MS_{\text{within}} = \frac{\sum_{j=1}^{k}\sum_{i=1}^{n_j}(X_{ij} - \bar{X}_j)^2}{n-k} \quad (12)$$

For the first data set of Table 4.3 MS_{between} and MS_{within} are obtained as follows:

$$\begin{aligned}
SS_{\text{between}} &= 3\times(4-4)^2+3\times(6-4)^2+3\times(2-4)^2=24\\
MS_{\text{between}} &= 24/2=12\\
SS_{\text{within}} &= (9-4)^2+(1-4)^2+(2-4)^2+\\
&\quad (10-6)^2+(2-6)^2+(6-6)^2+\\
&\quad (1-2)^2+(5-2)^2+(0-2)^2=84\\
MS_{\text{within}} &= 84/6=14
\end{aligned}$$

How can these variances be related to the two models mentioned before? Before answering to this question, we have to introduce a crucial concept, that of expected values, symbolized by $E(.)$; the dot indicates the variable whose expected value has to be determined. The expected value of a discrete variable X is given by the following formula:

$$E(X) = \sum(X_i \times p_i) \qquad (13)$$

where p_i stands for the respective probabilities of X_i; $E(X)$ is the value that is expected to be the average when a great number of X values are available. The traditional example is the expected value of a fair die; here $X_i = i$ $(i = 1, 2...6)$ and p_i is a constant: 1/6. $E(X) = \sum(i \times 1/6) = 3.5$. This outcome shows that the expected value is not necessarily a value one actually gets when throwing a die. The expected value equals the average of values observed when the 'experiment' is repeated a great number of times.

Back to our problem of relating the variances to the two models. If it is assumed that the within variance or error variance is the same in each group $(\sigma_1^2 = \sigma_2^2 = \cdots = \sigma_k^2 = \sigma_\varepsilon^2)$, the expected value of MS_{within}:

$$E(MS_{\text{within}}) = \sigma_\varepsilon^2 \qquad (14)$$

holds for both models. The expected value of MS_{between} depends on which model is true. The following expected values turn out to hold:

$$\begin{aligned} \text{if } H_0 \text{ is true}: E(MS_{\text{between}}) &= \sigma_\varepsilon^2 && (15) \\ \text{if } H_1 \text{ is true}: E(MS_{\text{between}}) &= \sigma_\varepsilon^2 + n\sigma_\alpha^2 && (16) \end{aligned}$$

n is the number of elements within a group (given equal cell frequencies). If H_0 is true, the division of s_b^2 (or MS_{between}) by s_w^2 (or MS_{within}) has an expected value of 1. If, however, H_1 is true, the expected value is higher than 1.

In order to clarify the concept of the expected value of mean squares, we would like to consider the actual sampling process in more detail. To that end we assume a population from which three samples are drawn; each sample consists of i participants. The population is characterized by a μ of 10, and a variance σ_ε^2 of 1.2. Let us assume that a very great number of experiments of the type discussed above, are carried out with three treatments applied to three samples drawn from the population $(\mu = 10, \sigma_\varepsilon^2 = 1.2)$. We assume that the treatments do not have any effect on the scores (H_0: $\alpha_1 = \alpha_2 = \alpha_3 = 0$). That is why we only deal with a single population. The situation can be described as follows:

population: $\mu = 10$, $\sigma_\varepsilon^2 = 1.2$			
sample:			
exp. 1	$\bar{X}_1\ \bar{X}_2\ \bar{X}_3 \rightarrow$	$MS_{\text{between}} = 1.0$	$MS_{\text{within}} = 0.9$
exp. 2	$\bar{X}_1\ \bar{X}_2\ \bar{X}_3 \rightarrow$	$MS_{\text{between}} = 0.8$	$MS_{\text{within}} = 1.1$
exp. 3	$\bar{X}_1\ \bar{X}_2\ \bar{X}_3 \rightarrow$	$MS_{\text{between}} = 1.3$	$MS_{\text{within}} = 1.2$
.	. . .	.	.
.	. . .	.	.
exp. N	. . .	.	.
(N = very large)		mean $MS_{\text{between}} = 1.2$	mean $MS_{\text{within}} = 1.2$

If the assumption that there are no effects holds, the means of both MS_{between} and MS_{within} over a very large number of samples (experiments) are equal to the error variance in the population, which is 1.2. In other words: $E(MS_{\text{between}}) = E(MS_{\text{within}}) = \sigma_\varepsilon^2$. If the assumption does not hold, however, the mean of MS_{between} exceeds the mean of MS_{within} (given that n is large). In this case the expected value of MS_{between} is $\sigma_e^2 + n\sigma_\alpha^2$.

In genuine experiments the means of the *MS*s are not available, but we know the distribution of the ratio of both mean squares under discussion if the H_0 is true, i.e. the F distribution. The F distribution permits us to determine to what extent the observed ratio of *MS*s is a probable result under the H_0. If the result is not 'probable enough', the H_0 is rejected and the alternative hypothesis is accepted.

- Decompose the deviation score $X_{ij} - \bar{X}$ into two components.
- What is the expected value of the random variable X, when it takes the values 1, 2, 3 and 4, and when $p_1 = p_2 = p_3 = p_4 = 0.25$?
- Give the model of the score X_{ij} in the presence of one effect and error.
- Is $MS_{\text{total}} = MS_{\text{between}} + MS_{\text{within}}$?
- Assume t tests are carried out on the scores of all pairs of four different samples; an α level of 0.01 is adopted. What is the probability of incorrectly rejecting at least one null hypothesis?

4.4. Testing effects: the F distribution

The statistical distribution of the quotient of two variances is known as the F distribution. This distribution is in fact a family of distributions. The form of the F distribution is a function of two elements: the degrees of freedom associated with the mean square in the numerator and the degrees of freedom associated with the mean square in the denominator, df_1 and df_2 respectively. The distribution can be computed if both mean squares or variance estimates have the same expected value. This is the case in our one-way analysis of variance if H_0 is true; both variances have the same expected value and the expected F value is 1.

In Figure 4.1 two F distributions are depicted which are associated with two pairs of degrees of freedom: (a) $df_1 = 4, df_2 = 8$, written as $F_{4.8}$, and (b) $df_1 = 4, df_2 = 30$, written as $F_{4.30}$.

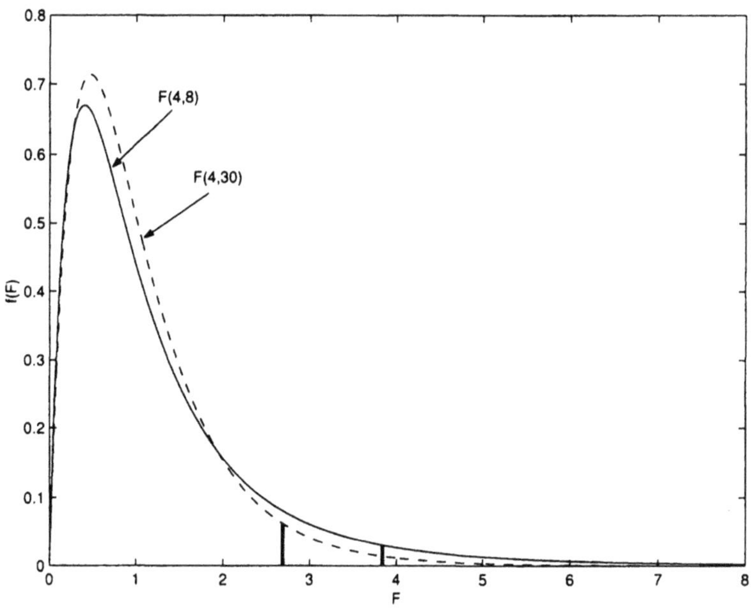

Figure 3.1. Plots of the probability density function of $F_{4.8}$ (= solid line) and $F_{4.30}$ (= dotted line); the corresponding critical values of F for $\alpha = 0.05$ are 3.84 and 2.69

Figure 4.1 illustrates that the form of the so-called probability density function of F depends on the associated degrees of freedom. In the example, the lower number of df_2 in $F_{4.8}$ results in a right tail that approximates the abscissa (x axis) more slowly than the right tail of $F_{4.30}$. If the α level is set at 0.05, the critical value of $F_{4.8}$ is 3.84, whereas it is much lower for $F_{4.30}$: 2.69. There are two entries for tables containing the critical values of the F distribution: df_1 and df_2. These two determine the value of the associated critical value. Such a table is given in Appendix C, where the critical values for two significance levels are reproduced: the 5% and the 1% level. In Table 4.4 we reproduce a section of the F table.

Table 4.4. A section of the F table; critical values of F, 5% and 1% (bold-face) points; df_1 = degrees of freedom numerator; df_2 = degrees of freedom denominator

	df_1											
df_2	1	2	3	4	5	6	7	8	9	10	11	12
1	161	200	216	225	230	234	237	239	241	242	243	244
	4052	**4999**	**5403**	**5625**	**5764**	**5859**	**5928**	**5981**	**6022**	**6056**	**6082**	**6106**
2	18.51	19.00	19.16	19.25	19.30	19.33	19.36	19.37	19.38	19.39	19.40	19.41
	98.49	**99.01**	**99.17**	**99.25**	**99.30**	**99.33**	**99.34**	**99.36**	**99.38**	**99.40**	**99.41**	**99.42**
3	10.13	9.55	9.28	9.12	9.01	8.94	8.88	8.84	8.81	8.78	8.76	8.74
	34.12	**30.81**	**29.46**	**28.71**	**28.24**	**27.91**	**27.67**	**27.49**	**27.34**	**27.23**	**27.13**	**27.05**
4	7.71	6.94	6.59	6.39	6.26	6.16	6.09	6.04	6.00	5.96	5.93	5.91
	21.20	**18.00**	**16.69**	**15.98**	**15.52**	**15.21**	**14.98**	**14.80**	**14.66**	**14.54**	**14.45**	**14.37**
5	6.61	5.79	5.41	5.19	5.05	4.95	4.88	4.82	4.78	4.74	4.70	4.68
	16.26	**13.27**	**12.06**	**11.39**	**10.97**	**10.67**	**10.45**	**10.27**	**10.15**	**10.05**	**9.96**	**9.89**
6	5.99	5.14	4.76	4.53	4.39	4.28	4.21	4.15	4.10	4.06	4.03	4.00
	13.74	**10.92**	**9.78**	**9.15**	**8.75**	**8.47**	**8.26**	**8.10**	**7.98**	**7.87**	**7.79**	**7.72**
7	5.59	4.74	4.35	4.12	3.97	3.87	3.79	3.73	3.68	3.63	3.60	3.57
	12.25	**9.55**	**8.45**	**7.85**	**7.46**	**7.19**	**7.00**	**6.84**	**6.71**	**6.62**	**6.54**	**6.47**
8	5.32	4.46	4.07	3.84	3.69	3.58	3.50	3.44	3.39	3.34	3.31	3.28
	11.26	**8.65**	**7.59**	**7.01**	**6.63**	**6.37**	**6.19**	**6.03**	**5.91**	**5.82**	**5.74**	**5.67**

In Table 4.4 we can find the critical value at 0.05 for $F_{4,8}$. We need to combine the column under $df_1 = 4$ with the row for $df_2 = 8$. The value found is 3.84. The critical value for $\alpha = 0.01$ is 7.01.

Most computer programs, however, do not give the critical values, but provide the probability of observing an F value equal to or greater than the observed F ratio when H_0 is true, i.e., when both mean squares have the same expected value. The derivation of the expected value of the MS_{between} is rather complex, especially if more than one factor is studied or if complications arise, for instance because participants have been measured more than once (repeated measures). The interested reader is referred to the standard introductions on analysis of variance (Winer, Brown, and Michels 1991, Kirk 1995). What is of interest here is that the between variance is increased by an additional component if H_0 is not true, and that this addition is caused by a group effect α_j. The decision strategy with respect to H_0 and H_1 is based on the value of the quotient of the within and between groups variance

($MS_{between}/MS_{within}$). If the value exceeds a critical value, H_0 is rejected (= model 1) and H_1 is accepted instead (= model 2).

This decision strategy can be illustrated by the outcomes of the analysis of the two data sets given in Table 4.1. As a matter of fact there are three possible SPSS approaches to the analysis of these data: a) General Linear Model, Univariate, b) Compare Means, One-Way ANOVA, and c) Mixed Models. General Linear Model, Univariate is the more general procedure, to be used when one or more independent variables ('factors') are included in the design; it has a lot of options. The Mixed Models approach is discussed at length in Chapter 9. For our example we use the One-Way ANOVA procedure. In Table 4.5 the relevant part of the SPSS output of that procedure for data set A is reproduced. Table 4.6 contains the same information for data set B.

Table 4.5. (Part of the) SPSS output from an analysis of variance (Compare Means, One-Way ANOVA) applied to data set A in Table 4.1

	Sum of Squares	df	Mean Square	F	Sig.
Between Groups	① 24.000	2	12.000	.857	.471
Within Groups	② 84.000	6	14.000	④	⑤
Total	③ 108.000	8			

In Table 4.5 the following observations can be made:

① The 'between' part relates to the effect α.

② The 'within' part represents the error (the 'within groups variance').

③ Adding up the sum of squares of both sources gives the total sum of squares; the same holds for the number of degrees of freedom.

④ The group effect is tested by dividing the mean square of the group effect α by the mean square of the residual part. The result is the F ratio.

⑤ The resulting F ratio in Table 4.5 is not significant (the ratio is even smaller than 1); H_0 (= model 1) has to be accepted.

The results in Table 4.6 are different:

① The F ratio is higher, i.e. 12.000.

② The resulting F ratio is significant at the 1% level; H_1 (= model 2) is accepted.

When an F ratio turns out to be significant, we have reason to reject the H_0 that both mean squares have the same expected value. For instance, if the F

Table 4.6. (Part of the) SPSS output from an analysis of variance (Compare Means, One-Way ANOVA) applied to data set B in Table 4.1

	Sum of Squares	df	Mean Square	F	Sig.
Between Groups	24.000	2	12.000	12.000	.008
Within Groups	6.000	6	1.000	①	②
Total	30.000	8			

value resulting from $MS_{\text{between}}/MS_{\text{within}}$ is significant, and if the expected values are $E(MS_{\text{between}}) = \sigma_\varepsilon^2 + n\sigma_\alpha^2$ and $E(MS_{\text{within}}) = \sigma_\varepsilon^2$, we may assume that $n\sigma_\alpha^2 \neq 0$. This assumption means that there is somewhere a difference between the levels of a factor. It does not mean that all levels of a factor have different effects! If there are more than two levels, it is quite possible that some of them do not differ from each other. If one wants to assess the difference between specific levels, so-called post hoc tests (also called a-posteriori or unplanned tests) have to be used. This procedure will be discussed in Section 4.6.

We have seen that $E(MS_{\text{between}}) = \sigma_\varepsilon^2 + n\sigma_\alpha^2$, and $E(MS_{\text{within}}) = \sigma_\varepsilon^2$. These expressions enable us to determine the estimator of σ_α^2, that is to say the variance of the effects, which is only relevant in case the group effect is a random factor. When we subtract the latter value from the former, we get the following result for Table 4.5: 12 – 1 = 11. This figure estimates the value of $n\sigma_\alpha^2$. As $n = 3$, σ_α^2 is estimated by 11/3 = 3.66. Using symbols like σ_α^2 implies that there is a population with a parameter σ_α^2, from which a number of levels was sampled. An example is given by an experiment in which a number of babies are tested in order to find out whether they are able to recognize faces or specific stimuli. While the population of babies is large, the number of babies actually included in the sample is relatively small. In this case the factor 'babies' is called a *random* factor: A relatively small number of levels is randomly sampled from the population. If the number of levels in the sample is approximately the same as the number of levels in the population, for instance we want to assess the effects of teaching methods, the factor is called *fixed*.

The distinction between random and fixed factors has far-reaching consequences for the expected values of the mean squares, and consequently for the pairs of mean squares which have to be used to calculate F ratios. Apart from the conceptual differences between a fixed and a random factor, there are also statistical differences involved. In most cases these only show up in the expected values of mean squares, but in some instances, where variance components are at issue, more detailed differences come to the surface. The main difference between a fixed factor A and a random factor A is that the effect α_j of the latter constitutes a random variable, with a variance σ_α^2, and an expected value $E(\alpha_j) = 0$, whereas the effect α_j of a fixed factor is a constant quantity, subject to the constraint that $\sum \alpha_j = 0$. These differences are the consequence of the assumption that the levels of a random factor are just a sample of a large population of levels. The sum of the corresponding effects in the experiment is not necessarily zero, but the expected value of an effect α_j is. As all possible levels of a fixed factor occur in the sample, the sum of the effects α_j cannot be anything other than zero, as an effect is defined as a deviation score: $\alpha_j = \mu - \mu_j$. These differences become manifest when we write out the expected value of the mean square associated with the treatment effect A of a simple one-way design, with the following score model:

$$X_{ij} = \mu + \alpha_j + \varepsilon_{ij} \qquad (17)$$

The expected values of the mean squares of the fixed and random factors respectively are:

$$\text{FIXED: } E(MS_A) = \sigma_\varepsilon^2 + n\sum \frac{\alpha_j^2}{k-1} \qquad (18)$$

$$\text{RANDOM: } E(MS_A) = \sigma_\varepsilon^2 + n\sigma_\alpha^2 \qquad (19)$$

The difference lies in the second term: For the random factor it is a genuine variance, whereas for the fixed factor the term contains a summation of squared effects. In many textbooks, and in statistical software, these terms are symbolized by σ^2, both for the random and the fixed factors. We adhere to this practice until Chapter 9.

4.5. One-way analysis of variance in SPSS

The relevant parts of the SPSS data matrix are reproduced in Table 4.7. The matrix consists of variables ('age' and 'accent') and cases (the participants).

Table 4.7. Relevant parts of the SPSS data matrix representing data set A from Table 4.1

age	accent
1	9
1	1
1	2
2	10
2	2
2	6
3	1
3	5
3	0

The following steps lead to an analysis of variance in SPSS:

window	Data View	
	click on	Analyze
	go to	Compare Means
	go to +	
	click on	One-Way ANOVA
window	Univariate	
the dependent variable:	*click on*	the variable 'accent'
	insert	under Dependent list
the independent variable:	*click on*	the variable 'age'
	insert	under Factor
	click on	OK

The output appears in an output file on the screen. An option within the Univariate window is called Random Factor(s). As explained in Chapter 2, SPSS can also run in syntax mode, where the commands have a text format. Window commands can be translated into syntax format by the Paste option. Our one-way example delivers the following syntax by clicking on Paste:

```
ONEWAY
  accent BY age
  /MISSING ANALYSIS
```

The first component, 'accent BY age', specifies the dependent variable which is split up into the groups defined by the variable 'age'. /MISSING refers to the fact that the analysis takes missing data into account if there are any. Syntax commands can be set to work by clicking on Run in the window of the syntax file.

4.6. Post hoc comparisons

As mentioned before, a significant F ratio simply leads to the rejection of the global ('omnibus') hypothesis that all effects or levels of a specific factor are equal to zero. We did not assess which specific effects were present. In other words, a significant F ratio very often calls for a further, more elaborate analysis. In section 4.2 we introduced analysis of variance as an alternative to carrying out a large number of comparisons between means obtained at the different levels of a factor. The probability of incorrectly rejecting at least one null hypothesis in k t tests (Type I error) is:

$$1 - p(\text{not rejecting } H_0)^k = 1 - (1 - \alpha)^k \qquad (20)$$

However, there are situations in which a researcher would like to know, after having established a significant F ratio in her or his analysis of variance, which means differ and which do not. Thus we appear to be back with the original problem, i.e. how does one control the inflation of the Type I error? (This Type I error is called the 'family-wise' error rate because all comparisons between means are relevant). The procedures which are used for this kind of analysis are called *post hoc comparisons*. As a matter of fact a large number of procedures is available; most of them are variants of the more general, and quite simple Bonferroni approach, which boils down to dividing the desired significance level α by the number of comparisons k. Thus, with a factor of four levels, $\alpha = 0.05$ becomes $0.05/12 = 0.0422$, as $(4 \times 3 \times 2)/2 = 12$ comparisons can be made. Statistical packages offer a number of alternatives, often as a function of 'Equal Variances Assumed' or 'Equal Variances Not Assumed'. (Un)equal variances refer to the variances within the groups. For 'Equal Variances Assumed' we mention:

- Bonferroni's adjustment
- The Newman-Keuls test
- Tukey's Honest Significant Difference test (HSD)
- Tukey's b test
- Kramer's modification of the HSD test

- Sidak's test
- Duncan's Multiple Range test
- Dunnett's test: (compares means with a single 'control' mean)
- Scheffe's test (this is known as the most conservative test)
- Hochberg's GT2 test
- Gabriel's pairwise Comparison test
- Ryan-Einot-Gabriel-Welsch (R-E-G-W) Multiple Stepdown Procedures
- Ryan-Einot-Gabriel-Welsch (R-E-G-W-Q) test based on the Studentized Range test
- Ryan-Einot-Gabriel-Welsch (R-E-G-W-F) test based on an F test.

For 'Equal Variances Not Assumed' the following procedures are available:

- The Games and Howel Pairwise Comparison test
- Tamhane's T2 test
- Tamhane's T3 test
- Dunnett's C test.

It is crucial to control the probability level of rejecting a true H_0 hypothesis when one makes a series of comparisons. In the section on one-way analysis of variance we explained how the application of a series of *t* tests brings about a high probability level of committing a Type I error.

How a statistical test for post hoc comparisons should be applied is illustrated with the one-factor data set in Table 4.8.

Table 4.8. Fictituous data set for post hoc comparisons: one factor, three areas

	urban areas		
	I	II	III
	5	7	5
	2	6	4
	4	7	6
	4	8	5
$\bar{X}_j$	3.75	7.00	5.00

After determining that the 'area' factor is significant at the 0.05 level ($F_{2.9}$ = 12.66), one may wish to know which areas do not differ from each other and should, therefore, be regarded as equivalent. In other words, we would like to compare areas I, II, and III pair-by-pair. We have to resort to procedures which aim at keeping α at an acceptable level. There are quite a few of these available. The literature on this topic suggests that Tukey's Honestly Significant Difference test (= HSD) and the Tukey-Kramer modification of it (TK) in the case of unequal sample *n*s are a good bet for most occasions.

In Tables 4.9 to 4.11 part of the output of the SPSS procedure is reproduced, in which a one-way analysis and associated post hoc comparisons were carried out for the data of Table 4.8. The analysis of variance results are presented in Table 4.9.

Table 4.9. (Part of the) SPSS output from an analysis of variance (Compare Means, One-Way ANOVA) applied to the data in Table 4.7

	Sum of Squares	df	Mean Square	F	Sig.
Between Groups	21.500	2	10.750	11.057	.004
Within Groups	8.750	9	.972		
Total	30.250	11			

The analysis of variance results in Table 4.9 show that the between groups factor ('urban areas') is significant. In the next step we applied the Tukey HSD test ($\alpha = .05$). The result of the multiple comparisons analysis according to the HSD test is given in Table 4.10.

Table 4.10. Post hoc comparisons (HSD test) for the data from Table 4.8

(I) Area	(J) Area		Mean Difference (I-J)	Std. Error	Sig.	95% Confidence Interval Lower Bound	Upper Bound
1	2	①	-3.2500*	.69722	.003	-5.1966	-1.3034
	3		-1.2500	.69722	.226	-3.1966	.6966
2	1	②	3.2500*	.69722	.003	1.3034	5.1966
	3	①	2.0000*	.69722	.044	.0534	3.9466
3	1		1.2500	.69722	.226	-.6966	3.1966
	2	②	-2.0000*	.69722	.044	-3.9466	-.05345

*. The mean difference is significant at the .05 level

Table 4.10 shows the following results:

① Pairwise comparisons between the mean scores obtained in urban 'area 1' and urban 'area 2', or urban 'area 2' and urban 'area 3' result in significant differences.

② These comparisons are redundant as the differences between areas 1 and 2, and 2 and 3 are the same as the differences between 2 and 1, and 3 and 2, apart from the sign.

These findings are summarized in the next table, in which the homogeneous sets are given.

Table 4.11. Output post hoc comparisons (Tukey HSD test) for data from Table 4.8

		Subset	
Urban Area	N	1	2
1	4	3.7500	
3	4	5.0000	
2	4		7.0000
Sig.		.226	.1000

Table 4.11 points to two homogeneous subsets: One contains the means of urban areas 1 and 3 and the other urban Area 2. Consequently, the first and third urban areas should be considered as equivalent. Because of the example discussed here one may get the impression that post hoc comparisons can only be carried out for one-way designs. This is not the case. Post hoc comparisons can be used in all kinds of designs; the crucial aspect is the selection of the appropriate error term. If comparisons have to be made between cells that vary along the levels of factors A and B, the error term should be the same as the one that was used to test the interaction AB. The data needed to calculate post hoc comparisons are: the magnitudes of the relevant means, the numbers of observations on which the cell means are based, the MS of the error term and its associated degrees of freedom. Post hoc comparisons can be defined as procedures to assess which treatments are considered to be as similar or dissimilar in their effects on the dependent variable. All treatments must be taken into account, and a 'yardstick' is calculated for pairwise comparisons. These procedures can in fact be seen as specific cases of a more general approach to the assessment of differences between treatment means: contrast analysis.

4.7. Determining sample size

Before an experiment is carried out in which the data are analyzed with analysis of variance we have to determine four elements, analogous to the list given in Chapter 3 for t tests:

1. the α level adopted
2. the size of the effect to be detected
3. the power of the test required to detect the effect on the basis of these data
4. the sample size needed.

To calculate the sample size needed to achieve agiven power tables can be used, like the ones given in Winer et al. (1991) and Kirk (1995) or dedicated software, like SAMPLEPOWER or NQUERY ADVISOR.

For two-sample tests it is quite easy to establish the effect size one wants to detect: a specified difference between the means of the subpopulations from wich the samples are drawn. SAMPLEPOWER uses f, after the index introduced by Cohen (1988), which takes the within-group variance into account. This index can be used to compare two means. We discussed this index in Chapter 2. For analysis of variance things are more complicated, as we have to deal with more than two subpopulations. The following index for effect size is in use:

$$f = \sqrt{\frac{\Sigma(\mu_j - \mu)^2/k}{\sigma_\varepsilon^2}} = \sqrt{\frac{SS_{\text{between}}}{\sigma_\varepsilon^2}} \qquad (21)$$

Of course a researcher does not think in terms of a-priori SS_{between} but rather in terms of differences between group means or the range of groups means he/she would like to detect. Fortunately programs like SAMPLEPOWER and NQUERY ADVISOR allow for this kind of input. Below we give an example based on Table 4.1, data set B. The calculations were carried out with NQUERY ADVISOR. Let us assume that we would like to detect differences between the means ranging from 2 to 6; we estimate that the within cell variation is 1 ($\sigma_\varepsilon = 1$). The power to be achieved is 0.80, and the significance level is set at 0.05. What is the sample size needed to achieve a power of 80%? Keeping in mind the data presented in Table 4.1, the input to the program was the following:

- number of levels of the factor = 3
- alpha level = 0.05
- power to be achieved = 0.80
- estimated within-cell variation = 1 (= σ_ε); the estimated within-cell variance = 1 (= σ_ε^2)
- range of means = 2 to 6.

With these data the index f for the effect size is calculated. In our case f = 1.633; the effect size squared (= Δ^2) is 2.667. This is the input needed to run the program. The results produced by NQUERY ADVISOR are given in Table 4.12.

Table 4.12. Output of NQUERY ADVISOR on needed sample size given a set of input parameters

Test significance, α	0.050
Number of groups, G	3
Variance of means, $V = \Sigma(\mu_j - \bar{\mu})^2/G$	2.667
Common standard deviation, σ	1.000
Effect size, $\Delta^2 = V/\sigma^2$	2.6670
Power (%)	80
n per group	3

Table 4.12 tells us that 3 cases per level are needed to achieve a power of minimally 0.80 (to be more precise, the actual power is 0.92; with two cases the power would be 0.54). As a matter of fact we achieved a significance level of 0.008 with 3 observations per cell.

4.8. Power post hoc

If a test fails to detect an effect, we have to report the power of the test. Statistical software is able to deliver the so-called *observed power* of a test. As is the case with t tests SPSS does not report 'observed power' for a one-way analysis. Again one has to resort to the General Linear Model procedure ('Univariate', with the option 'observed power') to obtain the power of the current analysis. For the purpose of 'power' just look at a part of the output in Table 4.13 under the heading 'group'. We analyzed the data set given in Table 4.1 (data set A). Below in Table 4.13 we reproduce part of the output obtained with the General Linear Model procedure.

Table 4.13 shows that the number of observations in each sample was too small to achieve the desired power. F was not significant, and the achieved power to detect differences between levels of the 'group' factor is only 0.139. The Table also gives values of the 'Noncent. Parameter', which refers to the noncentrality parameter d. Together with df_1 and df_2 this parameters determines the distribution of F if the effects α_i are not zero, as is the case when we test the H_0 that all α_is are zero. The magnitude of d depends on the value

Table 4.13. Part of the SPSS output for the General Linear Model, Univariate procedure applied to the data of Table 4.1 (data set A); $n = 3$ per cell

Source	df	Mean Square	F	Sig.	Noncent. Parameter	Observed Power(a)
Corrected Model	1	12.000	.857	.471	1.174	.139
Intercept	1	10.286	.018	.018	1.714	.139
Group	2	12.000	.857	.471	1.714	.139
Error	6	14.000				
Total	9					
Corrected Total	8					

a Computed using alpha = .05

of the sum of the squared treatment effects relative to the error variance. In post hoc power analysis we do not have to bother about this parameter. In fact, it is only of relevance when one uses tables to determine the power of an anlysis or the sample size to detect specified treatment effects.

4.9. Suggestions for statistical reporting

EXAMPLE 1
The mean scores obtained in the four groups of 10 observations each (with standard deviations in brackets) were: 7.1 (0.99), 8.4 (0.70), 6.1 (0.74) and 5.9 (0.74). A one-way analysis of variance was carried out on the data, and an α level of 0.05 was adopted. The 'group' factor was considered to be a fixed factor. 'group' was significant: $F_{3.36} = 20.403$, $p < 0.01$. Levene's test for homogeneity of variances resulted in $F_{3.36} = 0.396$, $p = 0.760$, ns. Thus homogeneity of variances was warranted. Subsequent post hoc comparisons (Tukey's HSD test), with an α level of 0.05 showed that groups 3 and 4 do not differ significantly (they form a homogeneous subset), while groups 1 and 2 are significantly different and differ from the aforementioned set of groups 1 and 2. (In Chapter 6 we discuss Levene's test which is used to assess the assumption of equal variances in the groups).

EXAMPLE 2 (ns = not significant)
The mean scores obtained in the four groups of 10 observations each were (with standard deviations in brackets): 7.1 (3.51), 8.4 (5.78), 6.1 (4.01) and 5.9 (3.14). A one-way analysis of variance was carried out on the data, and an α level of 0.05 was adopted. The 'Group' factor was not significant: $F_{3.36}$

= 0.731, $p = 0.541$, ns. Levene's test for homogeneity of variances resulted in $F_{3,36} = 0.318$, $p = 0.812$, ns. Therefore homogeneity of variances was warranted. The observed power to detect differences between the group means was 0.190, a low value, probably due to the relatively high variation within the groups.

4.10. Terms and concepts

- *Effect*: The influence of treatment j on the scores of the members of subpopulation j which undergo this treatment: $\alpha_j = \mu_j - \mu$.
- *F ratio*: A statistic formed by the quotient of two mean squares; this quotient is distributed as F when the expected values of both numerator and denominator are the same.
- *Factor*: An independent variable that influences the magnitude of scores.
- *Fixed factor*: A factor of which the levels represent the relevant categories in the population.
- *Level*: One of the treatments that make up a factor.
- *Mean squares*: The sum of squared deviations of scores (= sum of squares) from a specific quantity (group mean or other terms) divided by the appropriate degrees of freedom.
- *Random factor*: A factor whose levels are selected at random from a much larger population of levels.
- *Residual variance*: The variance that remains when part of the variance has been attributed to specific sources.
- *Sum of squares*: The sum of squared deviations of scores from a specific quantity (e.g. group mean).

4.11. Exercises

1. Determine whether the following F ratios are significant at the 5% level: $F_{10,20} = 2.80$; $F_{2,8} = 3.10$; $F_{2,80} = 3.10$; $F_{2,800} = 3.10$; $F_{5,100} = 2.25$; $F_{10,100} = 2.25$.
2. Fill in the missing data in the following analysis of variance table:

Source	*SS*	*df*	*MS*	*F*
Between	200	....	....	4
Within	....	....	12.5	
Total	....	44		

3. The following scores are given in a one-way table with three treatments:

A_1	A_2	A_3
1	4	7
2	5	8
3	6	9

   Carry out an analysis of variance by computer.

4. Calculate the MS_{between} and MS_{within} for the data contained in Table 4.8 by hand.
5. Data obtained in a one-factorial design (factor 'treatment' (3 levels)) are analyzed with a one-way analysis of variance. After having established a significant effect for the 'treatment' factor, post hoc comparisons are carried out. What is the maximum number of homogeneous sets that can be found?
6. What are the possible resulting model equations in a one-factorial design if the factor at issue is not significant?
7. Is the following equation correct: $MS_{\text{A}} + MS_{\text{error}} = MS_{\text{total}}$?
8. A one-way analysis of variance is carried out; the number of levels for the factor is 5, the number of observations per treatment 14. Determine the number of degrees of freedom for the F ratio to be used to test the factor treatment.

Chapter 5
Multifactorial designs

5.1. Preview

In this chapter we discuss designs which include more than one factor. The power of analysis of variance is the ability to deal with the effects of more than one factor or independent variable on a dependent variable. The effect of the combination of two or more independent variables (called *interaction*) is given a great deal of attention in Sections 5.2 and 5.6, as it plays an important role both in theory and practice. In Section 5.3 we introduce the distinction between *random* and *fixed* factors. This distinction is not only very important from a conceptual point of view, but it also plays a crucial role in finding the appropriate error terms for F ratios (Sections 5.4 and 5.5). In Section 5.7 the complete procedure for the analysis of a standard multifactorial design is summarized. In Section 5.8 we give an overview of a number of frequently used design types (*repeated measures designs* and *hierarchical designs*). Repeated measures designs are in a way similar to t tests for matched samples. In hierarchical designs not all levels of one factor are combined ('cross') with all levels of another factor. In Section 5.9 we discuss the concepts underlying hierarchical designs and the analysis of data obtained in these designs. Repeated measures designs, in which participants are tested in different conditions, are discussed separately in Chapter 8. Finally, in Section 5.10 a short introduction is given to the *analysis of covariance* (ANCOVA), a technique which is applied when an independent continuous variable supposingly affects the dependent variable.

5.2. Multifactorial designs and interaction

Thus far only a single-factor design (also called a one-way design) has been discussed. The power of analysis of variance, however, lies in its ability to investigate the effects of more than one independent variable on the dependent variable simultaneously. Suppose that the data given in Table 5.1 were sampled in a single-factor experiment with three treatments.

There seems to be a 'break' in the scores within the groups in Table 5.1. The first two participants achieved lower scores than the last two. By inspect-

Table 5.1. Single-factor design with three treatments

treatment		
I	II	III
1	3	9
2	4	10
4	8	14
5	9	15
3	6	12

ing their personal profiles, it might turn out that the first two members of each group were highschool graduates, whereas the last two graduated from university. Consequently, part of the variation within the groups might be due to an 'education' factor we did not take into account in the original model. The scores, originally displayed in a single- or one-factor design, can be rearranged in a two-factor design as shown in Table 5.2.

Table 5.2. Two-factor design for the data of Table 5.1

	treatment		
education	I	II	III
highschool	1	3	9
	2	4	10
university	4	8	14
	5	9	15

The rearranged data in Table 5.2 clearly suggests that the original one-factor design has to be replaced by a two-factor design. The scores seem sensitive to the effects of two factors: treatment and educational level. The model tested should be expanded by including an effect β_j for education:

$$X_{ijk} = \mu + \alpha_i + \beta_j + \varepsilon_{ijk} \qquad (1)$$

Apart from the question how such a model with two effects can be tested, an additional complication may arise when two factors are involved.

An important concept, not only in the context of analysis of variance, but also in other techniques is that of interaction. To illustrate this concept, the cell means of four data sets are given in Table 5.3, followed in Figure 5.1 by a graphical representation. In Table 5.3, the variable A is the row variable and variable B the column variable; the row variable receives the subscript i, the column variable the subscript j.

Table 5.3. Cell means of four data sets, showing interaction ((c) and (d)) and no interaction ((a) and (b))

data set (a): no interaction

	B_1	B_2	B_3
A_1	10	15	8
A_2	15	20	13

data set (b): no interaction

	B_1	B_2	B_3
A_1	8	13	18
A_2	11	16	21

data set (c): interaction

	B_1	B_2	B_3
A_1	7	10	13
A_2	9	14	25

data set (d): interaction

	B_1	B_2	B_3
A_1	8	11	14
A_2	16	13	9

All four data sets of Table 5.3 have two factors. Factor A has two levels, factor B three. The means of the six cells are displayed in Figure 5.1; the ordinate (y axis) relates to the cell values, the abscissa (x axis) to the levels of factor B.

What do the data and their associated graphs suggest? Interaction is said to occur in those cases where the differences between the levels of one factor are not equal for all levels of another factor. In data set (b), for instance, the difference between A_1 and A_2 is 3 at all levels of B: $11 - 8 = 3$, $16 - 13 = 3$, $21 - 18 = 3$. In data set (c) on the other hand, the differences between A_1 and A_2 are not the same at all levels of factor B: The difference at level B_1 is $9 - 7 = 2$, at B_2 $14 - 10 = 4$, and at B_3 $25 - 13 = 12$. In that case the two factors are said to *interact*. Likewise no interaction effect occurs in data set (a), whereas data set (d) shows a clear interactional pattern. Interactional patterns cannot be reproduced by an additive model equation in which only the two main effects α_i and β_j are included. A supplementary term has to be included in addition to the terms α_i and β_j.

The following model equation accounts for the mean cell values in the data sets (a) and (b):

$$\bar{X}_{ij.} = \mu + \alpha_i + \beta_j \qquad (2)$$

A triple subscript (X_{ijk}) is used. The first subscript indicates the row, the second the column, and the third the measurement within the cell. A dot notation

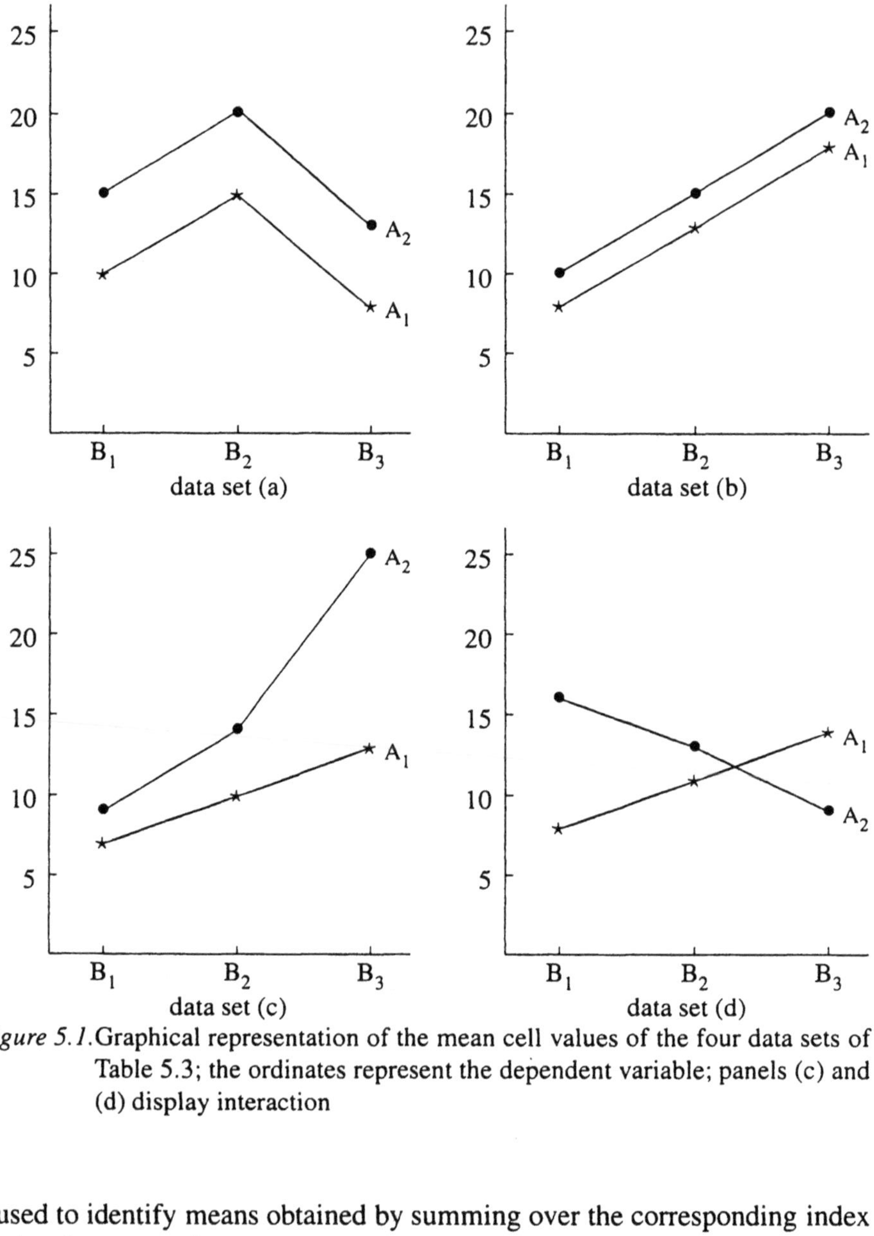

Figure 5.1. Graphical representation of the mean cell values of the four data sets of Table 5.3; the ordinates represent the dependent variable; panels (c) and (d) display interaction

is used to identify means obtained by summing over the corresponding index (in the above equation, the mean over the measurements within the cells).

For data set (b) the estimates of α_i and β_j are: $\alpha_1 = -1.5$, $\alpha_2 = 1.5$, and $\beta_1 = -5$, $\beta_2 = 0$, $\beta_3 = 5$. In fact, estimates should be written as $\hat{\alpha}_i$, $\hat{\beta}_j$ or a_i, b_j. We did not use this notation, because there is no danger of ambiguity between population parameters and their estimates. The estimate of μ is 14.5. Thus, for instance, score $\bar{X}_{12.}$ and $\bar{X}_{23.}$ can be partitioned as follows:

$$\begin{aligned} \bar{X}_{12.} &= 14.5 - 1.5 + 0 = 13 \\ \bar{X}_{23.} &= 14.5 + 1.5 + 5 = 21 \end{aligned}$$

It is not possible to construct such an additive model for data sets (c) and (d). A systematic difference emerges between the scores predicted by the model and the scores observed. An interaction term $(\alpha\beta)_{ij}$ has to be included to obtain a perfect fit for the means:

$$\bar{X}_{ij.} = \mu + \alpha_i + \beta_j + (\alpha\beta)_{ij} \qquad (3)$$

For data set (c), $\alpha_1 = -3$ and $\alpha_2 = 3$; $\beta_1 = -5$, $\beta_2 = -1$, and $\beta_3 = +6$. Given these values and the mean cell scores, the values of the interaction term can be determined. For instance:

$$\begin{aligned} \bar{X}_{12.} &= 13 - 3 - 1 + 1 &= 10 \\ \bar{X}_{23.} &= 13 + 3 + 6 + 3 &= 25 \end{aligned}$$

The terms $+1$ and $+3$ are specific to the cells (1,2) and (2,3) respectively. More precisely, they belong to specific combinations of levels i and j of the factors A and B. The interaction effect can be calculated as follows: The effects α_i and β_j denote the deviations of the subpopulation means from the overall population mean: $\alpha_i = \mu_{i.} - \mu$, $\beta_j = \mu_{.j} - \mu$. The interaction term $(\alpha\beta)_{ij}$ represents the deviation of the cell mean μ_{ij} from the overall mean minus the effects of α_i and β_j. This gives the following outcomes for the interaction effect:

$$\begin{aligned} (\alpha\beta)_{ij} &= \mu_{ij} - (\mu + \alpha_i + \beta_j) \\ &= \mu_{ij} - \mu - \alpha_i - \beta_j \\ &= \mu_{ij} - \mu - (\mu_{i.} - \mu) - (\mu_{.j} - \mu) \\ &= \mu_{ij} - \mu_{i.} - \mu_{.j} + \mu \qquad (4) \end{aligned}$$

In a two-factor design the number of measurements is $npq = N$. It is assumed that we have an equal number n of measurements in each cell; p is the number of rows (or number of levels of factor A); q is the number of columns (or number of levels of factor B). If $p = 2$, $q = 3$, and $n = 3$, the data can be represented as shown in Table 5.4.

Table 5.4. Two-factor design with $p = 2$, $q = 3$, and $n = 3$

	1	2	3	Row mean
1	X_{111} X_{112} X_{113}	X_{121} X_{122} X_{123}	X_{131} X_{132} X_{133}	$\bar{X}_{1..}$
2	X_{211} X_{212} X_{213}	X_{221} X_{222} X_{223}	X_{231} X_{232} X_{233}	$\bar{X}_{2..}$
Column mean	$\bar{X}_{.1.}$	$\bar{X}_{.2.}$	$\bar{X}_{.3.}$	$\bar{X}_{...}$

The mean square (*MS*) for the interaction effect can be calculated by dividing the interactional sum of squares (*SS*) by the appropriate number of degrees of freedom. Given the notational system of Table 5.4 and the equation found for the interaction effect, the *SS* can be calculated by the formula:

$$SS_{\text{interaction}} = \sum_{i=1}^{p} \sum_{j=1}^{q} (\bar{X}_{ij.} - \bar{X}_{i..} - \bar{X}_{.j.} + \bar{X}_{...})^2 \qquad (5)$$

The number of degrees of freedom associated with the interaction sum of squares is $(p-1)(q-1)$. The *MS* is the *SS* divided by the associated number of degrees of freedom. The remaining *SS* is called the 'error' or 'within cell variation'; the associated number of degrees of freedom is $pq(n-1)$.

5.3. Random and fixed factors

How the effects in a two-factor design are tested depends on the status of the factors involved. Factors can be *random* or *fixed*, and this distinction plays a crucial role in defining the appropriate mean square against which the mean square of an effect has to be tested. The distinction between random and fixed factors is based on the status of the categories or treatments in the sample, in relation to the categories or treatments in the population. The categories of the factor may represent the relevant categories in the population (the factor is *fixed*), or they may constitute a random sample of all categories in the population (the factor is *random*). The question as to whether all levels of a factor are included in the experiment is relevant. The answer has serious consequences for the calculation of the test statistic and even for the possibility of testing particular null hypotheses in more complex designs. When we test the efficiency of three methods of pronunciation training, we are primarily interested in the specific methods investigated, for instance A_1, A_2, and A_3. We are not interested in all possible methods, or the question whether different methods generally result in different pronunciation achievements. The methods A_1, A_2, and A_3 are chosen deliberately and in a second investigation the same methods are chosen again. These methods represent our population of 'levels' fully. In other cases, however, not all levels of a factor are included in the research design. An obvious example is the choice of 'words' which represent a word class category. When we test the reaction times of participants regarding two specific types of words, we choose, perhaps 'at random', words out of a dictionary or lexical data-base. The selected words do not constitute the whole population of possible words, just a subset. We would like to generalize our findings with regard to other words of the word class category in question. These two examples illustrate two types of factors, fixed and random:

- A factor is called *fixed* if all possible levels of that factor are included in the experiment ('methods' in our example).
- A factor is called *random* if the number of possible levels of that factor greatly exceeds that of the number of levels included in the experiment. Furthermore, the levels included should have been selected at random ('words' in the example given above).

The distinction between fixed and random factors might suggest that only the latter enable the researcher to generalize her/his findings to levels not included in the experiment. For example, when we 'use' students of a linguistics class as participants in an experiment, our sample is not based on a random selection procedure. Consequently, the participants do not constitute a random selection of all possible levels of the population. However, the question whether the findings may or may not be generalized to other participants, is not directly and uniquely related to the dichotomy of random and fixed factors. The question that has to be answered is whether the researcher has reason to assume that his participants constitute a sample which represents the population he is interested in. If the task in the experiment is not of a linguistic nature, he/she may argue that the participants are representative of the population of young people with a university background; generalization to that population may then be sufficiently warranted.

The absence or presence of fixed and random factors in a design leads to three types of models: a *fixed model* (containing only fixed factors), a *random model* (with random factors only), and a *mixed model* (with both random and fixed factors). The distinction between random and fixed factors has serious consequences for the expected values of the mean squares, and consequently for the pairs of mean squares which must be used for calculating F ratios. Apart from the conceptual differences between a fixed and a random factor, there are also statistical differences involved. In most cases these differences only show up in the expected values of mean squares, but in some instances, when variance components are at issue, more detailed differences surface. The main difference between a fixed factor A and a random factor A is that the effect α_j of the latter constitutes a random variable, with variance σ^2_α, and an expected value $E(\alpha_j) = 0$, whereas the effect α_j of a fixed factor is a constant quantity, subject to the constraint that $\sum \alpha_j = 0$. These differences are the consequence of the assumption that the levels of a random factor are simply a sample of a large population of levels. The sum of the corresponding effects in the experiment is not necessarily zero, only the expected value of an effect α_j is. As all possible levels of a fixed factor occur in the sample, the sum of the effects α_j cannot be anything other than zero, as an effect is defined as a deviation score: $\alpha_j = \mu - \mu_j$.

5.4. Testing effects in a two-factor design

We have shown how two factors and their interaction may affect the dependent variable, but we have not yet discussed how the presence or absence of factor effects can be tested statistically. To that end the expected values of the mean squares involved have to be derived. The expected value of a mean square - $E(MS)$ - is the mean of the sampling distribution of mean squares, obtained when repeated random samples are taken from the given (sub)population. In other words, it is the 'average' value of an *MS*. The derivation is quite a complicated matter and is not discussed here. An important element in this derivation is the question whether the factors have to be considered as 'random' or 'fixed'. The expected values in a two-factor design are given in Table 5.5.

In Table 5.5, n is the number of observations per cell, P and Q are the numbers of levels of factors A and B in the population, and p and q the respective numbers of levels actually included in the research design. The fractions $(P - p)/P$ and $(Q - q)/Q$ are 0 or (almost) 1 depending on the

Table 5.5. Expected values of mean squares in a two-factor design

Mean Square	df	Expected Value
MS_{A}	$p-1$	$\sigma_{\varepsilon}^2 + \frac{Q-q}{Q} n\sigma_{\alpha\beta}^2 + nq\sigma_{\alpha}^2$
MS_{B}	$q-1$	$\sigma_{\varepsilon}^2 + \frac{P-p}{P} n\sigma_{\alpha\beta}^2 + np\sigma_{\beta}^2$
MS_{AB}	$(p-1)(q-1)$	$\sigma_{\varepsilon}^2 + n\sigma_{\alpha\beta}^2$
MS_{error}	$pq(n-1)$	σ_{ε}^2

status of the levels of the factors. If p equals P, which means that factor A is fixed, the fraction is 0. If q is a random sample out of the population Q, the fraction is (approximately) 1. This implies that the choice of the error term in the F ratio depends on the status of the levels of the factors.

If factor A is a random factor, the appropriate error term to test the significance of factor B is the mean square of the interaction term. Since A is a random factor, it follows that $p \ll P$; consequently, $(P-p)/P = 1$, and the expected value of the mean square of factor B is $E(MS_{\mathrm{B}}) = \sigma_{\varepsilon}^2 + n\sigma_{\alpha\beta}^2 + np\sigma_{\beta}^2$. Note that a random effect has consequences for the expected values of other effects (additional components show up in the expected value); the mean square of the random factor remains the same (provided no other random effects are present). In testing factor B we have to find a mean square of which the expected value is the same as that of factor B, *minus* a term denoting the effect of factor B. In this case MS_{AB} meets this condition. The F ratio becomes:

$$F = \frac{MS_{\mathrm{B}}}{MS_{\mathrm{AB}}} \text{ with } df_1 = q-1,\ df_2 = (p-1)(q-1) \qquad (6)$$

In most research situations both factors are fixed. The error mean square then is the appropriate denominator in the F ratio.

The General Linear Model (= GLM), Univariate procedure in SPSS allows the user to specify whether a factor is fixed or random. Below we show how an analysis of variance can be carried out in SPSS on the data of Table 5.2. The relevant parts of the SPSS data matrix are reproduced in Table 5.6.

The data matrix in Table 5.6 shows how the two independent variables 'treat' and 'edu' should be coded. It is important to note that the order of the cases is not relevant for the outcome of the analyis. Any order will produce the same results in applying an analysis of variance.

Table 5.6. Relevant parts of the SPSS data matrix representing data in Table 5.2; 'treat' refers to the factor 'treatment' and 'edu' to the factor 'education'

treat	edu	dep
1	1	1
2	1	3
3	1	9
1	1	2
2	1	4
3	1	10
1	2	4
2	2	8
3	2	14
1	2	5
2	2	9
3	2	15

The following steps lead to a two-way analysis of variance in SPSS:

window Data View	*click on*	Analyze
	go to	General Linear Model
	go to +	
	click on	Univariate
window Univariate		
the dependent variable:	*click on*	the variable 'dep'
	insert	under dependent
the independent variables:	*click on*	the variable 'treat'
		the variable 'edu'
	insert	under fixed
	click on	OK

SPSS can also run in syntax mode, where the commands have a text format. Window commands can be translated into syntax format by the Paste option.

Our Univariate example delivers the following syntax which is automatically transferred to a syntax file by clicking on Paste:

```
UNIANOVA
  dep BY treat edu
  /METHOD = SSTYPE(3)
  /INTERCEPT = INCLUDE
  /CRITERIA = ALPHA(.05)
  /DESIGN = treat edu treat*edu .
```

The three variables involved are specified in the first line, 'dep' as the dependent variable, 'treat' and 'edu' as the independent variables. The design component indicates the two main effects and the interaction effect.

In Table 5.7 we give the output when an analysis of variance is carried out on the data in Table 5.2, assuming that both independent variables involved are fixed.

Table 5.7. (Part of the) SPSS output from an analysis of variance, the GLM, Univariate procedure, applied to the data in Table 5.2

Source ①	Type III Sum of Squares	df	Mean Square	F	Sig. ⑥
Corrected Model ②	227.000[a]	5	45.400	90.800	.000
Intercept ③	588.000	1	588.000	1176.000	.000
TREAT ⑥	168.000	2	84.000	168.000	.000
EDU	56.333	1	56.333	112.667	.000
TREAT * EDU	2.667	2	1.333	2.667	.148
Error ④	3.000	6	.500		
Total	818.000	12			
Corrected Total ⑤	230.000	11			

a. R Squared = .987 (Adjusted R Squared = .976) ⑦

Table 5.7 shows that the SPSS output distinguishes a number of sources of variation:

① The effects we implemented in our model equation; they are written in capitals: 'edu' (education), 'treat' (treatment) and their interaction: 'treat*edu'.

② Corrected Model: This source of variation is the total amount of variation associated with the effects incorporated in the model equation. In our example the sums of squares of the two main effects ('edu' and 'treat') plus the sum of squares of the two-way interaction 'treat*edu' give the *SS* of the 'Corrected Model' (sometimes called explained variation): $168.000 + 56.333 + 2.667 = 227.000$.

③ Intercept: The associated F ratio tests whether there is reason to assume that the cell means differ from 0 (intercept in terms of regression analysis, see Section 7.6); in most cases this source of variation is not relevant.

④ Error: Refers to the *SS* obtained by subtracting the *SS* of the Corrected Model from *SS* of the Corrected Total.

⑤ Corrected Total: The associated *SS* is obtained by summing the *SS* of the Corrected Model and that of Error.

⑥ In our example the two main effects ('edu' and 'treat') are significant at the 0.01 level, while the interaction 'treat*edu' is not significant ($p = 0.148$). This means that we do not have evidence that the effects of the factor 'treat' differ at the two levels of the factor 'edu'.

⑦ R is the correlation between the observed values and the values predicted by the model equation. That model equation contains a number of terms: the overall mean ('intercept'), the effects of 'treat', 'edu' and the interaction ('treat*edu'). Thus the model which predicts the scores X_{ijk} by X'_{ijk} is:

$$\boxed{X'_{ijk} = C + W_1\alpha_i + W_2\beta_j + W_3(\alpha\beta)_{ij} \qquad (7)}$$

The Greek symbols refer to the factors included in our model, and the Ws to the weights which have to be assigned to the factors. In Chapter 9 on unequal cells and regression analysis we will give more details about the way the factors and their levels can be coded. Of course, this model will not be able to predict the observed scores perfectly. The better the prediction is, the more acceptable the model. We can assess the predictive power of our model by correlating the observed scores with the scores obtained by the model. A high correlation implies a small magnitude of the error term ε_{ijk}. R^2 is the square of this correlation and expresses the percentage of the total variation accounted for by the model equation. R^2 is the sum of squares associated with the Corrected Model divided by the sum of squares of the Corrected Total (227.000/230.000=.987). If you want to use R^2 as an estimate of the percentage of the total variation in the population that can be explained, it needs to be adjusted for the sample size and the degrees of freedom as follows:

$$R^2_{adj} = R^2 - (1 - R^2)\frac{k}{N - k - 1} \qquad (8)$$

where N = sample size, and k = sum of the degrees of freedom of the effects involved. The total sum of the degrees of freedom in Table 5.7 is 5: 2 for 'treat', 1 for 'edu', and 2 for the interaction 'treat*edu'.

Plotting the interaction (which we leave to the reader) may suggest that the interaction 'treat*edu' is significant, as the difference between the mean scores of 'edu_1' and 'edu_2' is larger at 'treat_1' than at 'treat_2': 5 versus 3. The analysis of variance, however, does not support such a conclusion, as the associated p level is 0.148, which is not significant. The reader is invited to verify that effect 'edu' has to be tested against the interaction 'treat*edu' if factor 'treat' is a random factor (use the expected values given in Table 5.5).

5.5. Alternatives to testing effects in mixed designs

In Section 5.4 we showed how one can use expected values of *MS*s to find the error term in an F ratio to test a specific effect. We owe tables like Table 5.5 with expected values of Mean Squares in analysis of variance to – among others – Cornfield and Tukey (1956). This approach is extensively discussed (and followed up) in the standard textbooks of Winer et al. (1991) and Kirk (1995). However, it is not the only way of dealing with expected values of *MS*s. There has been, and still is, some debate on the expected value of MS_B in a mixed model where factor A is fixed and factor B random. In our table the interaction term in the expected value of the *MS* of the random factor is cancelled as $\frac{P-p}{P}$ is zero, because the number of levels of factor A in the population and in the sample are equal: $P - p = 0$.

Quite a number of statisticians disagree on the outcome of this procedure and assume that the interaction term, which reflects the effects of crossing a fixed factor A and a random factor B, is a normally distributed random variable $\sigma^2_{\alpha\beta}$, not restricted by any constraints, which should not be cancelled beforehand. The assumption that it has to cancelled is based on the constraint $\sum \alpha\beta_{ij} = 0$, which is analogous to the fixed effect assumption that $\sum \beta_j = 0$ (cf. Section 4.4). SPSS follows the assumption that $\sigma^2_{\alpha\beta}$ should not be cancelled and consequently sets the $E(MS_A)$ to $\sigma^2_\varepsilon + n\sigma^2_{\alpha\beta} + np\sigma^2_\beta$. This leads to the use of MS_{AB} as error term both for the fixed and the random factor.

This controversy can be solved, at least to some extent, by following the generally accepted procedure in multifactorial designs, which tells us that the interaction between main effects should be assessed first, before tests of main effects become relevant. Thus, after having accepted the hypothesis that $\sigma^2_{\alpha\beta} = 0$, we decide that $E(MS_{\text{B}}) = \sigma^2_{\varepsilon} + np\sigma^2_{\beta}$. Consequently the error term to test factor B becomes MS_{E}. Of course, there is one possible flaw in this procedure: the power of the test that $\sigma^2_{\alpha\beta} = 0$. If the power is low, we still do not know with certainty that $H_0 = 0$ and still have to decide whether MS_{AB} or MS_{error} should be used as error term. In general we can say that using MS_{AB} as error in a two-factorial design can be seen as a more conservative and consequently more secure procedure to test the hypothesis that the σ associated with the random factor is zero.

5.6. The interpretation of interactions

Something more should be said about the role of interaction in interpreting the data. Interactional patterns may take various shapes. A crucial distinction is made between *ordinal* and *disordinal* patterns. A disordinal (also called a *cross-over*) interaction is marked by a crossing of the lines which connect the mean cell values as shown in the graph for data set (d) in Figure 5.1. If there is an interaction effect, one often wants to know whether cell values at separate levels of a factor differ significantly. For instance, for data set (c) in Figure 5.1 one may want to know whether the two mean cell values at B_1 are significantly different. A specific kind of test can be applied to the differences between A_1 and A_2 at the separate levels of factor B. Such tests are called *simple main effects* tests. They will be discussed later (cf. Chapter 6).

An ordinal interaction pattern is marked by converging or diverging lines, whereas the differences between the mean cell values at at least one level of a factor remain substantial. This type of interaction is often the result of the way in which the dependent variable is measured. In many cases it is possible to 'remove' an ordinal interaction effect by a transformation of the dependent variable (cf. Chapter 7). A fictitious example may illustrate this phenomenon. Suppose that the second-language proficiency of participants is measured by the number of words they translate in a time interval of 100 seconds. Two groups of participants with a different first language background (B_1 and B_2) are tested and these two groups are further distinguished by the amount of time they have spent on learning the second language in question (S_1 represents the lower group and S_2 the higher one). The test performance of the participants can be expressed by the mean time they need to translate a word or by the total number of words translated during the test. The fictitious results are given in Table 5.8.

Table 5.8. Fictitious mean cell values on a translation task with two factors; B stands for language background, S for amount of study

language background	amount of time	number of words	mean time per word
B_1	S_1	5	20.00 sec
B_1	S_2	10	10.00 sec
B_2	S_1	10	10.00 sec
B_2	S_2	15	6.67 sec

The fictitious results of Table 5.8 are displayed in Figure 5.2. The graphs in this figure, for the mean time needed to translate a word and the number of words translated, show two different patterns. The graph for the mean time has two converging lines which points to the presence of an interaction effect. The graph for the number of words has two parallel lines; there is no interaction effect.

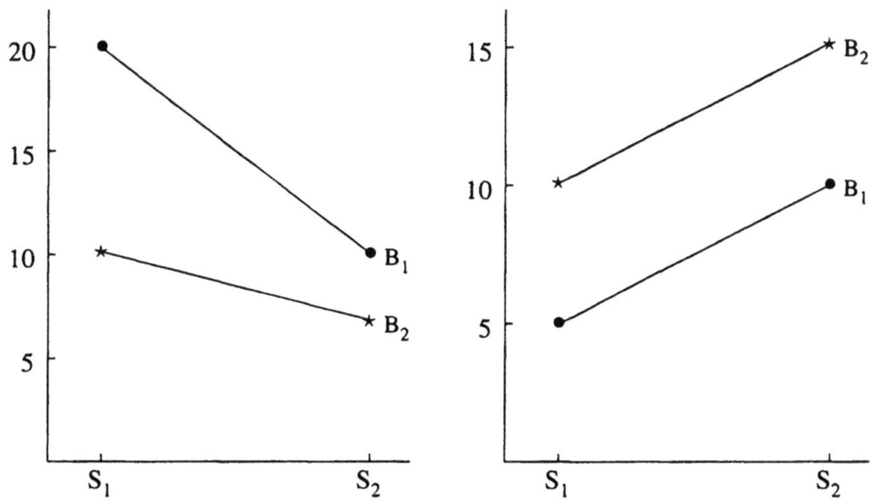

Figure 5.2. Graphical representation of the mean cell values of the data of Table 5.8; in the graph on the left the mean time is displayed on the *y* axis, in the graph on the right the number of words. B stands for language background, S for amount of study

The graph representing the mean time suggests that participants belonging to group B_1 profit more from the time spent on learning the second language; the graph representing the number of words suggests that there is no difference between the two groups with regard to the speed with which they have learned the second language. Apparently, the emergence of ordinal interaction depends on the measurement scale used and on whether or not the pattern is sensitive to a monotonic scale transformation. The researcher should ask which of the two representations of the dependent variable is the most appropriate. The examples of interaction given (cf. also Section 5.6) make clear that the occurrence of interaction affects the meaning of the main effects. The presence or absence of interactions play an essential role in the interpretation of effects in analysis of variance. In order to illustrate the role of interactions two examples will be discussed. Each respectively shows that (a) it is possible that interaction effects are present in the absence of any main effect, (b) and that it is not possible to qualify the main effects without specifying the interaction effects if both interaction and main effects are significant.

The data set discussed next illustrates the fact that interaction effects may occur in the absence of a main effect. Suppose that we have collected the data given in Table 5.9, which represent a two-way design with the two-level factors A and B.

Table 5.9. Fictitious data in a two-factor design with factor A (two levels) and factor B (two levels)

	B_1	B_2
	1	7
A_1	2	8
	3	9
	7	1
A_2	8	2
	9	3

The mean scores of the cells in Table 5.9 can be represented in a graphic way as shown in Figure 5.3. Figure 5.3 indicates that there are differences between the mean cell values. At the same time it is clear that no global or general main effect for any of the two factors is present. An analysis of variance carried out on these data yields the results given in Table 5.10.

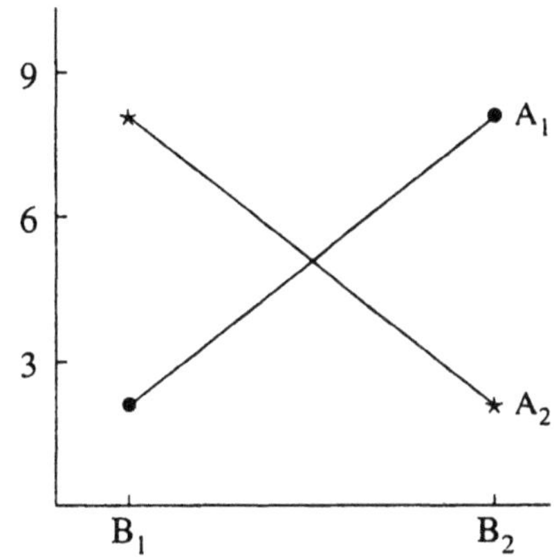

Figure 5.3. Plot of the interaction between factor A and B for the data of Table 5.9; the *y* axis represents the dependent variable

Table 5.10. (Part of the) SPSS output from an analysis of variance (GLM, Univariate) applied to the data in Table 5.9

Source	Type III Sum of Squares	df	Mean Square	F	Sig.
Corrected Model	108.000	3	36.000	36.000	.000
Intercept	300.000	1	300.000	300.000	.000
A	.000	1	.000	.000	1.000
B	.000	1	.000	.000	1.000
A * B	108.000	1	108.000	108.000	.000
Error	8.000	8	1.000		
Total	416.000	12			
Corrected Total	116.000	11			

a. R Squared = .931 (Adjusted R Squared = .905)

There is no significant main effect in Table 5.10! The disordinal interaction pattern is such that the mean values in levels 1 and 2 of the two factors level out; in the absence of differences between the levels the sums of squares of the main effects are zero. We now arrive at point (b) with respect to the role of interactions in interpreting main effects: If both interaction effects and main effects are significant, we cannot talk of global main effects any longer without specifying the interaction effects. In Figure 5.3 an interaction pattern is displayed, which indicates a substantial difference between A_1 and A_2 at level B_2, but no difference between the two A levels at B_1.

Suppose that the corresponding analysis of variance table – not reproduced here – yields significant main and interaction effects. If one says that

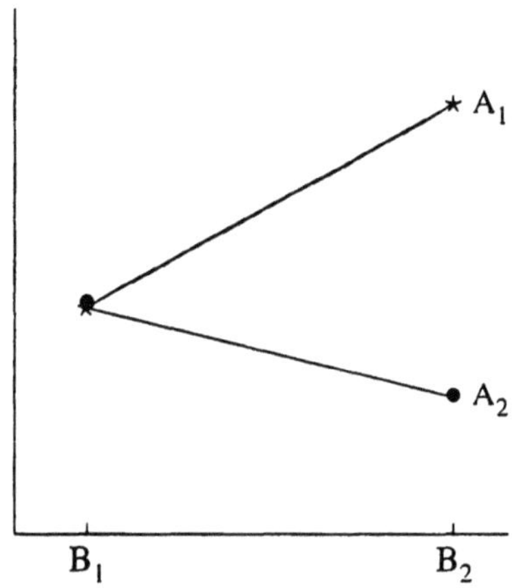

Figure 5.4. Plot of the interaction between factor A and B with no difference between the two A levels at B_1

effect A is significant without mentioning the interaction, one suggests that the magnitude of the dependent variable at A_1 and A_2 always differs significantly. This is not the case. The difference can only be found at level 2 of factor B.

The number of possible interaction effects grows exponentially if more factors are incorporated in the research design. In the kind of design discussed here, the number of possible interaction effects can be derived by simply listing all factor combinations. Three factors A, B and C, for instance, yield four possible interaction effects: $(\alpha\beta)$, $(\alpha\gamma)$, $(\beta\gamma)$, and $(\alpha\beta\gamma)$. The emergence of significant interaction effects of more than two variables often makes the interpretation of the results rather difficult.

In Figure 5.5 we display a possible pattern of interactions when the three-way interaction ABC is found to be significant. The panels in Figure 5.5 exhibit different AB interactions at the two levels of factor C: Consequently a three-way interaction ABC shows up. Figure 5.5 illustrates how the plotting of the interactional patterns may be helpful in interpreting more complicated interaction effects.

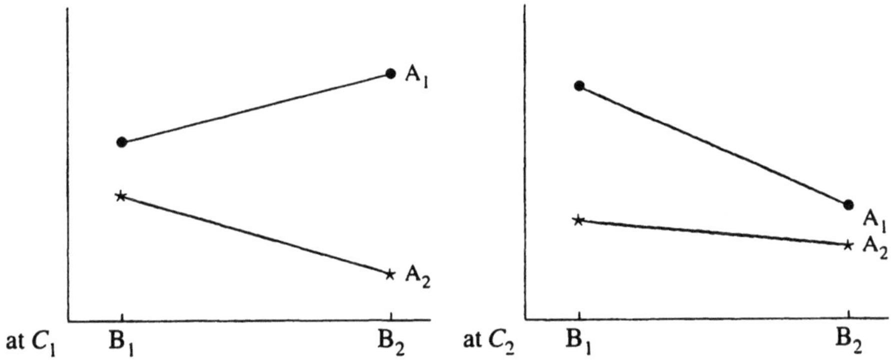

Figure 5.5. Plot of a three-way interaction between the factors A, B and C

5.7. Summary of the procedure

The successive steps of analysis of variance can be summarized as follows:

- Formulate a model equation with all the factors and their interactions; as an example we give the model for a completely randomized two-factorial design:

$$\boxed{X_{ijk} = \mu + \alpha_i + \beta_j + (\alpha\beta)_{ij} + \varepsilon_{ijk}}$$

- Decide whether the factors have to be regarded as fixed or random.
- Determine the expected values of the mean squares, for instance:

$$\boxed{E(MS_{\mathrm{B}}) = \sigma_\varepsilon^2 + n\sigma_{\alpha\beta}^2 + np\sigma_\beta^2}$$

  The expected value of MS_{B} implies that factor A is a random factor.
- Seek pairs of mean squares whose expected values are equal except for one term; this term constitutes the effect that has to be tested. Given the preceding expected value for factor B, the mean square needed is:

$$\boxed{E(MS_{\mathrm{AB}}) = \sigma_\varepsilon^2 + n\sigma_{\alpha\beta}^2}$$

- Calculate the sums of squares and the mean squares.
- Calculate the quotient of the relevant pairs of mean squares, for instance:

$$F = \frac{MS_{\mathrm{B}}}{MS_{\mathrm{AB}}} \quad \text{with } df_1 = q-1,\ df_2 = (p-1)(q-1)$$

- Determine the significance level (.05 for instance).
- Compare the observed F ratio with the critical F value; if the latter is smaller, reject the null hypothesis that the effect tested is zero.

> – Explain in your own words what is meant by 'interaction' between two factors.
> – What is the difference between an ordinal and a disordinal interaction pattern?
> – Is $F_{2,8} = 4.40$ significant at the 5% level (the subscript of F gives the degrees of freedom of the numerator and denominator)?
> – Express the interaction term $(\alpha\beta)_{ij}$ in terms of population means.
> – Does it make sense to consider a two-tailed F test?
> – What is the difference between random and fixed factors?

5.8. Other design types

Our introduction to analysis of variance was based on a specific type of design, i.e. the so-called *completely randomized factorial design*. This design is marked by two characteristics: (1) The participants are selected randomly within the groups or they are randomly assigned to the different treatments. Consequently, every subject is measured only once. (2) If more than one factor is involved, all its levels are *crossed* with all the levels of another factor; thus every level of one factor is combined with all the levels of the other factors. *Randomized block designs*, including designs with *repeated measures*, differ with respect to the first characteristic mentioned. Such designs will be discussed in Chapter 7. *Hierarchical designs* differ with respect to the second characteristic. They are marked by the presence of *nested* factors. A factor is called nested if the levels of that factor or treatment occur only within specific levels of another factor or treatment. The distinction between nested and crossed factors is illustrated in Table 5.11.

Suppose one wants to compare two Dutch speaking language communities, the Netherlands and Flanders (in Belgium), A_1 and A_2. Within a language community two dialect regions are selected. Thus there are four dialect regions in all: B_1 to B_4. Each dialect region belongs to one of the two language communities. This means that the factor B is nested within A. When one compares age groups, however, there is no problem selecting the same

Table 5.11. Comparison of a design with nested and a design with crossed factors

A_1		A_2	
B_1	B_2	B_3	B_4

Factor B
nested within A

A_1		A_2	
B_1	B_2	B_1	B_2

Factors A and
B crossed

two age groups for each language community, B_1 and B_2. The factors A and B are crossed. Hierarchical designs frequently occur in language research. For a hierarchical design, in which factor B is nested within A, the model equation has the following form:

$$X_{ijk} = \mu + \alpha_j + \beta_{k(j)} + \varepsilon_{ijk} \qquad (9)$$

The subscript $k(j)$ refers to the factor with subscript k which is nested within the factor with subscript j. In this model equation no interaction effect is present, because any specific effect of factor β is restricted to a specific value of factor α. No interaction effect can occur. The strength of this design is that part of the variation can be attributed to the nested factor. If this effect were not taken into account, the effect of the nested factor would be incorporated in the error term. This could lead to an unjustified overestimation of the strength of the error effect.

Table 5.12 gives the expected values of the mean squares of a hierarchical design. The expected values given hold for the configurations occurring most frequently.

Table 5.12. Expected values of mean squares in a hierarchical design with B as a random effect nested within A as a fixed effect

hierarchical design		
MS	*df*	Expected Value
MS_A	$p-1$	$\sigma_\varepsilon^2 + n\sigma_\beta^2 + nq\sigma_\alpha^2$
$MS_{B(A)}$	$p(q_{(j)}-1)$	$\sigma_\varepsilon^2 + n\sigma_\beta^2$
MS_{error}	$pq_{(j)}(n-1)$	σ_ε^2

Another important design is the *Latin square*, which is marked by the fact that each level of a factor C occurs only once under the levels of factors A and B. An example of a Latin square design is given in Table 5.13. Like the other basic design types mentioned in this chapter it can be seen as a building block in more complicated designs. The expected values of this design type are not discussed here. In SPSS one needs syntax to carry out an analysis of data obtained in a *Latin square*.

Table 5.13. Layout of a Latin square design; each level of factor C occurs only once under the levels of factor A and B

	A_1	A_2	A_3
B_1	C_1	C_2	C_3
B_2	C_2	C_3	C_1
B_3	C_3	C_1	C_2

The design of an experiment has consequences for the calculation of the mean squares, the degrees of freedom, the expected values of the mean squares and the F ratios. It is important, therefore, to be able to specify the nature of the design to be analyzed *exactly*. There is a wide variety of designs in which data can be collected and analyzed. The way more complicated designs can be worked out and analyzed is explained in standard textbooks, such as Winer et al. (1991) and Kirk (1995).

5.9. A hierarchical three-factor design

In this section we would like to discuss an example of a more complicated design based on a relatively limited data set. Let us assume that we want to investigate the effects of two teaching methods on a proficiency score. Six schools are involved, three of them randomly assigned to method B_1 and the other three to B_2. From each school five boys and five girls are selected randomly. The design, therefore, consists of the factors A (gender, two levels), B (method, two levels) and C (schools, six levels, nested within B). The data are given in Table 5.14.

The full model equation, containing all possible effects, has the following form:

$$X_{ijkl} = \mu + \alpha_i + \beta_j + \gamma_{k(j)} + (\alpha\beta)_{ij} + (\alpha\gamma)_{ik(j)} + \varepsilon_{ijkl} \qquad (10)$$

Because of the nesting of C within B, not all imaginable interaction effects are present. Only the crossed factors can produce an interaction effect: A with B, and A with C.

Table 5.14. Three-factor design with A (gender, fixed effect, two levels), B (method, fixed effect, two levels), and C (schools, random effect, six levels) nested under B

	B_1			B_2		
	C_1	C_2	C_3	C_4	C_5	C_6
	5	4	2	7	8	7
	4	3	4	5	7	7
A_1	3	2	3	6	6	4
	2	3	4	4	5	6
	4	2	3	5	6	3
	5	4	5	9	7	9
	4	3	4	8	8	8
A_2	2	4	3	7	8	9
	3	2	2	8	9	9
	4	4	3	9	8	9

An analysis of variance was carried out by means of the GLM, Univariate procedure of SPSS. The factors A and B are assumed to be fixed, whereas C is random. The data matrix has four variables. The three independent variables 'a', 'b', 'c', and the dependent variable 'dep'. In Table 5.14, the index of the variable 'c' goes from 1 to 6. The values in SPSS need to be coded differently. The indexing of 'b' restarts under each level of variable 'b'. The values of 4, 5, and 6 are coded as 1,2 and 3 in the SPSS data matrix.

The analysis of variance cannot be carried out with a point-and-click (PAC) procedure of SPSS. The univariate window does not offer the option of nesting. The SPSS syntax window must be used. The syntax is as follows:

```
UNIANOVA   dep BY a b c
  /RANDOM = c
  /METHOD = SSTYPE(3)
  /INTERCEPT = INCLUDE
  /CRITERIA = ALPHA(.05)
  /DESIGN = a, b, c WITHIN b,
  b BY a, a BY c WITHIN b .
```

The random nature of factor 'c' has to be mentioned explicitly here. The nesting of 'c' within 'b' has to be indicated in the design specification. The design specification lists all possible effects. Table 5.15 displays the results.

① In the ANOVA results in Table 5.15, the term MS_{Error} below the SPSS

Table 5.15. (Part of the) SPSS output from an analysis of variance (GLM UNIVARIATE) applied to the data in Table 5.14

Source		Type III Sum of Squares	df	Mean Square	F	Sig.
Intercept	Hypothesis	1612.017	1	1612.017	3022.531	.000
	Error	2.133	4	.533[a]		
A ③	Hypothesis	30.817	1	30.817	26.414	.007
	Error	4.667	4	1.167[b]		
B ④	Hypothesis	205.350	1	205.350	385.031	.000
	Error	2.133	4	.533[a]		
C(B) ⑥	Hypothesis	2.133	4	.533	.457	.766
②	Error	4.667	4	1.167[b]		
A * B ⑦	Hypothesis	20.417	1	20.417	17.500	.014
	Error	4.667	4	1.167[b]		
A * C(B)	Hypothesis	4.667	4	① 1.167	1.045	.394
	Error	53.600	48	② 1.117[c]		

a. MS(C(B)) ⑤
b. MS(A * C(B)) ②
c. MS(Error) ①

table refers to the within-cells variation. This error term is not suitable to test all effects. It is only used to test effect A*C(B). The B between brackets in C(B) indicates that B is nested within C. The other effects are tested for their significance with other error components.

② For A, A*B and C(B) the error term required is $MS_{\text{A*C(B)}}$. For the main effect B and the intercept it is $MS_{\text{C(B)}}$. To come to this conclusion, the composition of the associated expected mean square of the effects needs to be determined. In most cases we do not need these expected values. However they provide some insight into the principles of analysis of variance: They show how effects should be tested. Details on expected values of mean squares are given by Winer et al. (1991) and Kirk (1995).

③ Factor A (gender) turns out to be significant (at the 5% level). The F ratio, obtained by dividing MS_{A} by the mean square of the interaction AC with C nested within B (= $MS_{\text{AC(B)}}$), is 30.8167/1.1667 = 26.414.

④ Factor B (method) is also significant, as is the interaction AB.

⑤ The error term for B is the mean square of $C(B)$, for AB it is the mean square of AC(B). The interaction AB is displayed in Figure 5.6.
Figure 5.6 shows that the difference between levels B_1 and B_2 is not equal at both levels of factor A. This inequality is strong enough to bring about a significant interaction effect.

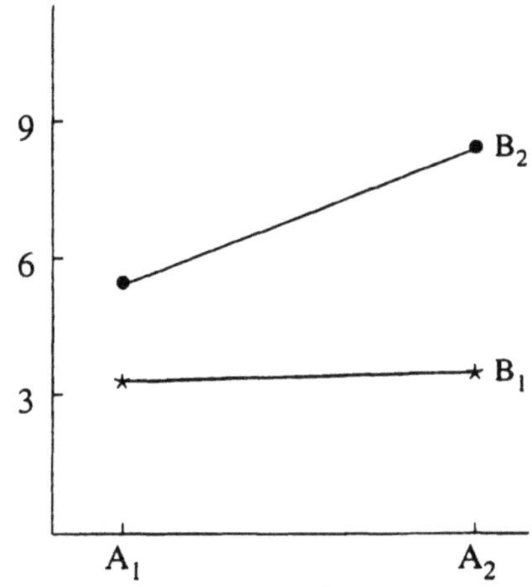

Figure 5.6. Interaction plot of the factors A and B in Table 2.20

⑥ Neither factor C(B) nor [⑦] the interaction AC(B) produce a significant result. Notice that the mean is also tested; in our example the resulting *F* ratio is significant (see 'intercept'). The hypothesis tested is that the mean value of all observations is zero. A significant result means that the mean observed is not zero. In most cases, therefore, testing the mean does not make much sense.

5.10. Analysis of covariance

Heterogeneity of participants in the cells of a variance-analytic design decreases the power of the analysis, as the error term gets inflated by the random variation. There is also variation which is not random; it can even be predicted – to some extent – by a variable which is relevant for the dependent variable. A variable like 'age' is a good example. Many performance related variables correlate with this factor: reading tests, comprehension, spelling etc. Children's achievements in this domain tend to increase as a function of age. There are two statistical ways in which such variables can be controlled. Either we introduce such a variable as a factor in our design (the factor then has a limited number of levels, for instance, very young, young, adolescent), or we enter it as a continuous variable in the model equation and regard it as a component of the error variance which can be removed. In the latter case the variable is called a *covariate*. In Figure 5.7 we depect a situation in which there is one dependent variable (X), one factor ('group') and one covariate (Y).

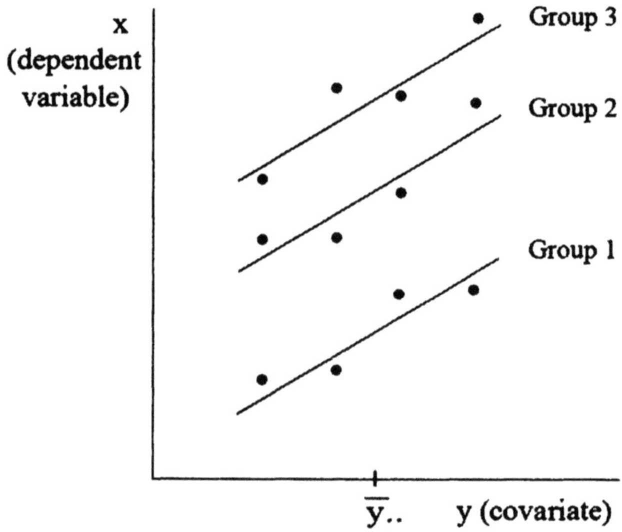

Figure 5.7. Equality of treatment-level regressions as assumed in the ANCOVA model. The levels refer to the 'group' factor, X to the dependent variable and Y to the covariate

Figure 5.7 shows the following:

- Apparently the mean values of the scores obtained in the three groups are different.
- The within-group variance is considerable; the scores on the dependent variable are correlated with scores on the covariate.
- Within each group there is a linear regression of the covariate on the dependent variable.
- The slopes of the regression lines are identical.

We could take advantage of this situation in an attempt to reduce the error variance. Figure 5.7 shows that the scores obtained on the dependent variable can be predicted well by a regression equation of the kind $\beta(Y_{ij} - \bar{Y}_{..})$ in which the deviation of a score of subject i in group j from the mean of the covariate Y predicts the score obtained on the dependent variable to a large extent (apart from the effects of the treatments, here: groups). The magnitude of the effect is symbolized by the slope β (a multiplication factor) of the line which relates X with Y.

When we rewrite the model equation for a one-way design and include the regression term mentioned above, we get:

$$X_{ij} = \mu + \alpha_j + \beta(Y_{ij} - \bar{Y}_{..}) + \varepsilon_{ij} \qquad (11)$$

in which the variable Y_{ij} denotes the value observed of covariate Y (in our example 'age') in subject i of group j, $\bar{Y}_{..}$ the overall mean of the covariate, and β the pooled (!) within-groups slope. Thus differences in scores which can be accounted for by differences in the covariate, can be discarded, as the original error term in the model equation of the one-way design now consists of two parts:

$$\varepsilon_{ij} = \varepsilon^*_{ij} + \beta(Y_{ij} - \bar{Y}_{..}) \qquad (12)$$

in which ε^*_{ij} is smaller than the original ε_{ij} of the one-way design in which the effects of the covariate Y are not taken into account. The new model with the covariate indicates that values on the dependent variable X also vary as a function of the covariate Y. There are two important assumptions for the covariance model (based on Figure 5.7):

1. Within each group, the dependent variable has a linear relationship with the covariate(s).
2. The slope of the regression line for each covariate is the same in each group. This assumption is called equal treatment-level slopes, as is depicted in Figure 5.7.

In Table 5.16, we show the data of a one-way design with the factor 'group' (3 levels) and the covariate 'age'.

Table 5.16. Pronunciation scores obtained in three groups of learners of a second language 'group', with 'age' as covariate

group 1		group 2		group 3	
age	score	age	score	age	score
20	70.0	22	80.0	22	95
23	60.0	26	70.0	33	85
27	45.0	29	65.0	37	80
29	55.0	40	60.0	41	75
35	45.0	44	55.0	48	70

We can carry out a covariance analysis with the PAC procedure. The syntax needed for a covariance analysis with 'age' as covariate is:

```
UNIANOVA dep BY group WITH age
  /METHOD = SSTYPE(3)
  /INTERCEPT = INCLUDE
  /CRITERIA = ALPHA(.05)
  /DESIGN = group age.
```

The data in Table 5.16 suggest that scores obtained in a test on L2 pronunciation decrease with older students (a negative correlation). In Tables 5.17a, b and c we present parts of the SPSS output for the analysis of the data given in Table 5.16. The first table gives the ANOVA table for the ordinary univariate analysis of this one-way design. The factor 'group' is significant: $F_{2.12} = 8.588, p = 0.005$.

Table 5.17a. Conventional one-way analysis applied to the data of Table 5.16

	Type III Sum of Squares	df	Mean Square	F	Sig.
GROUP	1703.333	2	851.667	8.588	.005
Error	1190.000	12	99.167		

Table 5.17b. Results of the covariance analysis of the data of Table 5.16 with 'age' as covariate; the GLM, Univariate procedure of SPSS was used

	Type III Sum of Squares	df	Mean Square	F	Sig.
AGE	977.323	1	977.322	50.549	.000
GROUP	2591.534	2	1295.767	67.019	.000
Error	212.677	11	19.334		

covariate	B	Beta	Std. Err.	t-value	Sig. of t
AGE	-1.06578	-.65203	.150	-7.110	.000

As can be seen in Table 5.17b, the effect of including 'age' as covariate leads to a substantial reduction of the *SS* of the error, from 1190.000 to 212.677. Consequently, the F ratio for GROUP increases.

The significant t value associated with the linear regression weight *Beta* ($t = -7.110$, $p = 0.000$) indicates that the dependent variable has a linear relationship with the covariate. This does not mean, however, that the first assumption of covariance analysis is met: A linear relationship between the dependent variable and the covariate within each group. Therefore we carried out linear regression analyses within each group. They all appeared to be (almost) significant: group 1: $F_{1,3} = 7.11$ ($p = 0.076$), group 2: $F_{1,3} = 24.47$ ($p = 0.016$) and group 3: $F_{1,3} = 398.96$ ($p = 0.000$). In light of the very small power of the test (only 4 observations in each group) we take these results as warrants for the assumption of linear relations between the covariate and the dependent variable in each group.

We also have to test the second assumption, viz. whether the slopes of the regression of the dependent variable with the covariate in the three groups are equal. This is assumed to be the case in the model which incorporates a covariate (cf. β in equation (12)). To that end we carry out an analysis of variance in which the interaction between the factor 'group' and the covariate is tested. If the interaction is *not* significant we can assume that the slopes are equal, and that the model equation is correct. The simpliest way to carry out that test is to use the syntax of the SPSS MANOVA procedure, (cf. Page, Braver, and MacKinnon 2003: 121-123) which goes as follows:

```
MANOVA dep BY group(1,3) WITH age
  /ANALYSIS = dep
  /DESIGN = group, age, group BY age.
```

The ANALYSIS subcommand overrides what is specified on the MANOVA command itself; thus a 'normal' analysis of covariance is not carried out. The results are presented in Table 5.17c. The only relevant output is the interaction term, which tests the homogeneity of regression slopes.

Table 5.17c. Test to assess whether the covariate 'age' interacts with the factor 'group'; the MANOVA procedure of SPSS was used

	SS	DF	MS	F	Sig of F
WITHIN+RESIDUAL	176.71	9	19.63		
GROUP	69.99	2	35.00	1.78	.223
AGE	939.99	1	939.99	47.87	.000
GROUP BY AGE	35.96	2	17.98	.92	.434

The interaction is not significant ($p = 0.434$), which means that we can accept the assumption of equal slopes.

5.11. Suggestions for statistical reporting

EXAMPLE 1
An analysis of variance (completely randomized two-factorial design) was carried out on the data. The two fixed factors were SES (3 levels) and Method (2 levels). Each cell of this design contained 10 observations. The significance level was set at 0.01. A significant interaction was found between SES and Method: $F_{2,54} = 53.904$, $p < 0.01$. The graph shows that the difference between levels of the factor SES differs as a function of the Method used to optimize reading skills. The factor SES was also significant at the 1% level: $F_{2,54} = 64.79$, while the factor Method was not significant: $F_{1,54} = 0.936$; the observed power for the latter test was 0.158.

EXAMPLE 2
The degree of 'accentedness' was established in four towns in two regions. In each town 30 informants were interviewed and their speech was subsequently rated by a panel of listeners. An analysis of variance carried out on the ratings, with the factor Regions (fixed) and the factor Towns (fixed, nested within Regions) yielded the following results: Regions: $F_{2,6} = 15.84$, $p < 0.01$, Towns: $F_{6,232} = 54.23$, $p < 0.01$ (In Chapter 6 we will discuss measures which are used to assess the strength of effects; they should be reported too).

EXAMPLE 3
An analysis of variance was carried out on the data. The random factor Hospitals was nested within the fixed factor Denomination (Protestant, Catholic and State hospitals). Each level of the latter factor contained three hospitals. In each hospital 10 nurses were randomly sampled. The factor Hospital was not significant at the 5% level ($F_{6,81} = 1.80$), whereas factor Denomination was ($F_{2,81} = 4.70$, $p < 0.05$).

5.12. Terms and concepts

- *Analysis of covariance*: Analysis of variance in which the effects of a specific variable (the 'covariate') on the dependent variable are taken into account. The covariate is not part of the design, but tends to inflate the error variance if not incorporated in the analysis.
- *Completely randomized factorial design*: A design in which every participant is measured only once and in which all factors are crossed. All partic-

ipants are randomly assigned to the cells of the design.

- *Fixed factor*: A factor whose levels represent the relevant categories in the population.
- *Hierarchical design*: A design containing nested factors.
- *Interaction*: Effect resulting from a specific combination of levels of factors. In a two-factor design the interaction effect $\alpha\beta_{ij}$ is $\mu_{ij} - \mu_j - \mu_i + \mu$. An important distinction is made between ordinal and disordinal interactions.
- *Ordinal interaction*: Interaction in which the direction of the line linking the mean scores of level i of factor A over the levels of factor B is the same as the line linking the mean scores of level j of factor A.
- *Disordinal interaction*: Interaction in which the direction of the line linking the mean scores of level i of factor A over the levels of factor B is the opposite of the line linking the mean scores of level j of factor A.
- *Nesting*: Factor A is said to be nested within factor B when not all levels of factor A are combined with all levels of factor B.
- *Random factor*: A factor whose levels are selected at random from a much larger population of levels.

5.13. Exercises

1. Suppose the following cell means and marginal means are found in a two-factor design:

	A_1	A_2	A_3	
B_1	4	8	12	8
B_2	8	16	24	16
	6	12	18	12

   What would the cell means be if no interaction were present?

2. A speech pathologist investigates the Voice Onset Time (VOT) of three groups of speakers: (1) severe stutterers, (2) moderate stutterers, and (3) people who do not stutter. The dependent variable is the VOT in /b/. Each group consists of 10 speakers. Give the design of this experiment. Are the factors crossed or nested?

3. In Table 5.7 an analysis of variance table is given, based on two factors: 'edu' and 'treat'. Determine the F value associated with factor 'edu' when factor 'treat' is regarded as random; use Table 5.5.
4. Give the model equation of a two-factorial design in which factor B is nested within factor A.
5. In which of the two-factorial designs is the interaction AB not relevant: (1) A and B crossed, (2) A nested within B?
6. Is it possible that in a two-factorial design both MS_A and MS_B are zero, whereas MS_{AB} is not zero?
7. The following data were obtained in a completely randomized factorial design:

	B_1	B_2
A_1	3	5
	4	6
	2	4
A_2	2	6
	3	6
	3	3

a) Carry out an analysis of variance by computer. Both factors are fixed. Report the F and p values.
b) Give the model equation before the ANOVA is carried out, and after.

8. The data obtained in a two-factorial hierarchical design with random factor B ('school') nested within factor A ('type of school') are given in the following table:

	A_1			A_2	
B_1	B_2	B_3	B_4	B_5	B_6
4	3	2	7	6	6
3	5	6	8	7	8
5	8	3	9	8	9
6	4	5	8	7	8

a) Carry out an ANOVA on these data and report the F and p values.
b) Show the SPSS syntax.

9. Stutterers were randomly assigned to three different therapies. The scores obtained in a final test administered after therapy are given in the following table (T2), together with the stutter scores registered before therapy (T1). It was decided to carry out a covariance analysis, with the scores obtained at T1 as covariate.

therapy 1		therapy 2		therapy 3	
T1	T2	T1	T2	T1	T2
30	30.0	30	25.0	25	25
40	35.0	40	20.0	30	25
45	45.0	30	20.0	40	40
30	30.0	40	25.0	40	35
35	30.0	35	20.0	45	40

Determine
a) whether the factor 'therapy' is significant at the 5% level;
b) whether the assumption of equal regression slopes of the covariate with the dependent variable is warranted.

Chapter 6
Additional tests and indices in analysis of variance

6.1. Preview

In Chapters 4 and 5 the fundamental concepts of analysis of variance were presented and illustrated on the basis of clear-cut examples. However, after having carried out a global analysis (sometimes called an 'omnibus analysis') a researcher may need more detailed information on the groups included in the design. Examples are provided by questions like: "If there is an ordinal interaction between factors A and B, do the mean values in cells i and $i+1$ of factor A differ significantly at level k of factor B?". This chapter gives an overview of techniques to answer these questions, i.e. *simple main effects analysis*, *post hoc comparisons* and *contrast analysis*, in Sections 6.2, 6.3 and 6.4 respectively. Other questions have to do with the magnitude of effects which are significant. Significance of an effect is not only the result of *effect size*, but also of the degrees of freedom and the error variance of the factors involved. A number of indices is known to reflect the magnitude of effects; the use of these indices will be discussed and illustrated ('strength of association') in Section 6.5. Finally, in Section 6.6 we give an overview of items which have to be mentioned when analysis of variance is reported.

6.2. Simple main effects

Stating that two factors interact is one thing, knowing and interpreting the full pattern of differences at the various levels of these factors is another. In Table 6.1 we present a simple two-factorial completely randomized design, with factors A (two levels) and B (two levels). The analysis of variance results for the data of Table 6.1 are given in Table 6.2.

The interaction AB in Table 6.2 is significant at the 5% level; we would like to know whether factor B is significant at both levels of factor A. In other words, we want to evaluate two supplementary null hypotheses:

H_0 :	$\mathrm{B}_1 = \mathrm{B}_2$ at level A_1
H_0 :	$\mathrm{B}_1 = \mathrm{B}_2$ at level A_2

Table 6.1. Data in a completely randomized two-factorial design

	B_1	B_2
	3	8
A_1	3	9
	2	7
	2	3
A_2	3	2
	6	5

Table 6.2. ANOVA table for the data of Table 6.1

Source	Type III Sum of Squares	df	Mean Square	F	Sig.
Corrected Model	52.917[a]	3	17.639	8.819	.006
Intercept	234.083	1	234.083	117.042	.000
A	10.083	1	10.083	5.042	.055
B	18.750	1	18.750	9.375	.016
A * B	24.083	1	24.083	12.042	.008
Error	16.000	8	2.000		
Total	303.000	12			
Corrected Total	68.917	11			

a. R Squared = .768 (Adjusted R Squared = .681)

The mean scores of the cells in Table 6.1 are represented in Figure 6.1. This figure shows a bigger difference between B_1 and B_2 at A_1 than at A_2.

A simple main effect analysis can be carried out with the MANOVA procedure in SPSS, which is only available in syntax mode; for further details, see Page, Sanford & MacKinnon (2003). The abbreviation MANOVA stands for Multivariate ANalysis Of VAriance. The syntax for our example is:

```
MANOVA dep BY a(1,2) b(1,2)
  /DESIGN = b WITHIN a(1) b WITHIN a(2) .
```

To illustrate the testing of simple main effects, part of the output of the SPSS MANOVA procedure is given in Table 6.3. Table 6.3 shows the following:

① Factor B has a significant effect at level 1 of factor A: $F_{1,9}$ for B within A(1) = 14.72, $p = 0.004$.

② Factor B does not have a significant effect at level 2 of factor A: $F_{1.9}$ for B within A(2) = 0.06, $p = 0.816$.

Notice an interesting property: Adding up the *SS*s of the simple main ef-

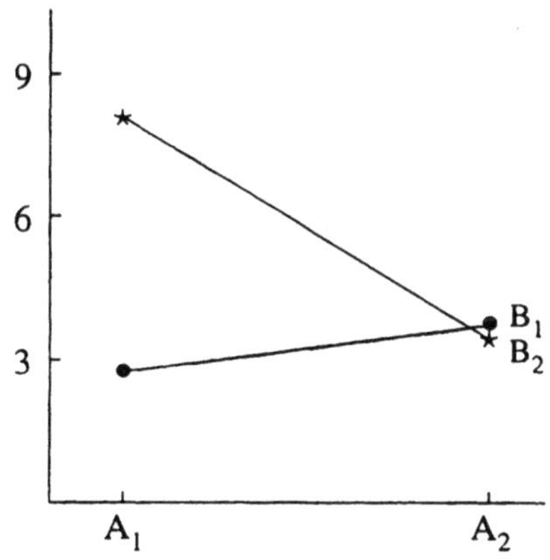

Figure 6.1. Plot of the interaction between factor A and B for the data of Table 6.1; the y axis represents the dependent variable

Table 6.3. (Part of) the SPSS output of the MANOVA procedure for the data of Table 6.1 to test simple main effects

Source of variation	SS	df	MS	F	Sig of F
WITHIN+RESIDUAL	26.08	9	2.90		
b WITHIN a(1) ①	42.67	1	42.67	14.72	.004
b WITHIN a(2) ②	.17	1	.17	.06	.816

fects results in 42.67 + 0.17 = 42.84. This sum equals the sum of *SS*s associated with factor B and the interaction AB given in the output of Table 6.2: $\sum B$ at $A_i = SS_B + SS_{AB}$. The question which error term should be used is not always easy to answer. If the factors at issue (here factor B and the interaction AB) have the same error term (MS_{within} in Table 6.3) there are no problems: In this case the simple main effects are tested against this error term (default in SPSS).

6.3. Post hoc comparisons in multifactorial designs

As mentioned before, a significant F ratio only leads to the rejection of the global hypothesis that all effects or levels of a specific factor are equal to zero. We have not assessed which specific effects are present. In other words, a significant F ratio very often calls for a further, more elaborate analysis.

As shown in Chapter 4 for one-way analysis, we can also carry out a post hoc comparisons analysis for a multifactorial design. This analysis can be performed both by PAC and by the use of the syntax option.

The following steps lead to an analysis of variance in SPSS for the data presented in Table 5.2, with two factors 'treat' en 'edu', plus a posthoc analysis for the factor 'treat':

window Data View	*click on*	Analyze
	go to	General Linear Model
	go to +	
	click on	Univariate
the dependent variable:	*click on*	the variable 'dep'
	insert	under Dependent list
the independent variable:	*click on*	the variable 'treat' and 'edu'
	insert	under Fixed Factor(s)
post hoc comparisons	*click on*	Post Hoc
	insert	the variable 'treat' under post hoc test
	click on	Tukey
	click on	continue
	click on	OK

The output will appear on the screen in an output file. The output of the post hoc comparisons is given in Tables 6.4 and 6.5. Table 6.4 gives the pairwise comparisons. Table 6.5 gives information on possible subsets,

Table 6.4. Part 1 of the output of the procedure Post Hoc Multiple Comparisons (Tukey's HSD test) for the data from Table 5.2

					95% Confidende Interval	
(I)TREAT	(J)TREAT	Mean Difference (I-J)	Std. Error	Sig.	Lower Bound	Upper Bound
1	2	-3.0000*	.50000	.002	-4.5341	-1.4659
	3	-9.0000	.50000	.000	-10.5341	-7.4659
2	1	3.0000*	.50000	.002	1.4659	4.5341
	3	-6.0000*	.50000	.000	-7.5341	-4.4659
3	1	9.0000*	.50000	.000	7.4659	10.5341
	2	6.0000*	.50000	.000	-4.4659	7.53415

Table 6.5. Part 2 of the output of the procedure Post Hoc Multiple Comparisons (Tukey's HSD test) for the data from Table 5.2

		Subset		
Treatment	N	1	2	3
1	4	3.0000		
2	4		6.0000	
3	4			12.0000
Sig.		1.000	1.000	1.000

As explained earlier, SPSS can also run in syntax mode, where the commands have a text format. Window commands can be translated into syntax format by the Paste option. Our two-factorial example delivers the following syntax by clicking on Paste:

```
UNIANOVA dep BY treat edu
  /METHOD = SSTYPE(3)
  /INTERCEPT = INCLUDE
  /POSTHOC = treat(TUKEY)
  /CRITERIA = ALPHA(.05)
  /DESIGN = treat edu treat*edu.
```

Post hoc comparisons can be used for all sorts of comparisons, including comparisons for parts of multi-factorial designs. The crucial aspect is the selection of the appropriate error term. The data needed to calculate post hoc comparions are: the magnitude of the relevant cell means, the numbers of observations on which the cell means are based, the *MS* of the error term used and its associated degrees of freedom.

SPSS has a syntax option that can be used for post hoc comparisons for parts of multifactorial designs. If one wants to carry out specific post hoc comparisons, for example on the means of the treatment factor in Table 5.2 at level 1 of the education factor, one needs to specify the following components:

- the values of the cell means at issue
- the number of observations on which the cell means are based (in our example 2)
- the value of MS_{error}, in our case 0.500
- the *df* associated with MS_{error}, in our case 6.

The syntax defines a dataset, followed by the commands LIST and ONEWAY:

```
MATRIX DATA VARIABLES =INDEP ROWTYPE_ dep /FACTOR=treat.
BEGIN DATA.
1 N 2
2 N 2
3 N 2
1 MEAN 1.50
2 MEAN 3.50
3 MEAN 9.50
. MSE 0.500
. DFE 6
END DATA.
LIST.
ONEWAY dep BY treat/MATRIX=IN(*)/RANGES=TUKEY.
```

Running this syntax file gives the standard output of the post hoc comparisons test, again for Tukey. The outcomes of Tukey's HSD test are given in Table 6.6.

Table 6.6. Part 1 of the output of the post hoc comparisons procedure (Tukey's HSD test), for the three cells of level 1 of the variable 'education' in Table 5.2

(I) treat	(J) treat	Mean Difference (I-J)	Std. Error	Sig.	95% Confidence Interval	
					Lower Bound	Upper Bound
1	2 ①	-2.0000*	.7071068	.067	-4.169598	-.169598
	3 ②	-8.0000	.7071068	.000	-10.169598	-5.830402
2	1	2.0000*	.7071068	.067	.169589	4.169598
	3 ②	-6.0000*	.7071068	.000	-8.169598	-3.830402
3	1	8.0000*	.7071068	.000	5.830402	10.169598
	2	6.0000*	.7071068	.000	3.830402	8.169598

Table 6.6 shows the following:

① At the first level of the variable 'education', treatments 1 and 2 constitute a homogeneous set.

② Treatments 1 and 2 differ significantly from treatment 3.

6.4. Contrasts

Post hoc comparisons can in fact be seen as specific cases of a more general approach to the assessment of differences between treatment means: contrast analysis. A contrast between k means is a linear combination of weighted means, in which the weights, c_j, add up to zero:

$$\psi = c_1\bar{X}_1 + c_2\bar{X}_2 + \cdots + c_k\bar{X}_k \qquad (1)$$

with $\sum c_j = 0$ (ψ is a Greek letter, pronounced as psi). If only two means are involved, the next contrast can be defined:

$$\psi = (+1)\bar{X}_1 + (-1)\bar{X}_2 = \bar{X}_1 - \bar{X}_2 \qquad (2)$$

This contrast, which is tested for being zero, is already familiar to us: It is equivalent to the two-sample t test, where this contrast appears in the numerator of the corresponding formula. The concept of 'contrasts' provides us with a powerful device to test patterns of differences between means. An example will illustrate this. Assume we have four treatments, assigned to four groups of subjects. The data are presented in Table 6.7.

Table 6.7. Fictitious data set for analysis by contrasts: one factor, four treatments

	urban areas			
	I	II	III	IV
	6	8	7	2
	5	7	6	3
	4	6	5	1
	3	5	5	2
$\bar{X}_j$	4.50	6.50	5.75	2.00

We would like to know whether treatments 1 and 2 differ as a group from treatments 3 and 4. The following contrast is set up:

$$\psi = (+\frac{1}{2})\bar{X}_1 + (+\frac{1}{2})\bar{X}_2 + (-\frac{1}{2})\bar{X}_3 + (-\frac{1}{2})\bar{X}_4 \qquad (3)$$

The coefficients $+\frac{1}{2}, +\frac{1}{2}, -\frac{1}{2}, -\frac{1}{2}$ add up to zero and distinguish two groups of means: $(\bar{X}_1 + \bar{X}_2)$ versus $(\bar{X}_3 + \bar{X}_4)$ by using the contrast $\frac{1}{2}(\bar{X}_1 + \bar{X}_2) - \frac{1}{2}(\bar{X}_3 + \bar{X}_4)$.

Other contrasts are also possible, for instance:

$$\psi = (+1)\bar{X}_1 + (-\frac{1}{3})\bar{X}_2 + (-\frac{1}{3})\bar{X}_3 + (-\frac{1}{3})\bar{X}_4 \quad (4)$$

Here group 1 is compared to (or contrasted with) the mean of groups 2, 3, and 4 by

$$\bar{X}_1 - \frac{1}{3}(\bar{X}_2 + \bar{X}_3 + \bar{X}_4). \quad (5)$$

It is obvious that a large number of contrasts between k groups can be defined. However, only a set of $k-1$ contrasts can be defined which are mutually independent, that is, without deriving them from other contrasts in the same set by addition or subtraction. For instance, the following set of three contrasts contains dependent contrasts:

$$\begin{array}{lcrrrr} \psi_1 & = & 0 & -1 & +1 & 0 \\ \psi_2 & = & -1 & 0 & +1 & 0 \\ \psi_3 & = & -1/2 & -1/2 & +1 & 0 \end{array}$$

The third (ψ_3) can be obtained by summing up the first and second ($\psi_1 + \psi_2$) and dividing the result by 2. Special attention should be paid to so-called orthogonal contrasts. *Orthogonality* means that the sum of the products of corresponding coefficients is zero. For instance the following two contrasts are orthogonal:

$$\begin{array}{lcrrrr} \psi_1 & = & +1 & -1/2 & -1/2 & 0 \\ \psi_2 & = & -1/3 & -1/3 & -1/3 & +1 \end{array}$$

as $(1 \times -1/3) + (-1/2 \times -1/3) + (-1/2 \times -1/3) + (0 \times +1) = 0$. Orthogonal contrasts use non-overlapping information, a property which facilitates the interpretation of the outcomes. The following set is an example of two nonorthogonal contrasts:

$$\begin{array}{lcrrrr} \psi_1 & = & -1 & +1/2 & +1/2 & 0 \\ \psi_2 & = & -1/2 & -1/2 & +1 & 0 \end{array}$$

In the first contrast the mean of $\bar{X}_2 + \bar{X}_3$ is compared with $\bar{X}_1$, whereas $(\bar{X}_1 + \bar{X}_2)/2$ and $\bar{X}_3$ are compared in the second. The subsets in both contrasts clearly overlap. That is why the sum of the products of corresponding coefficients is not zero, but 3/4.

An example of three orthogonal contrasts is the following:

ψ_1	=	+1	−1/2	−1/2	0
ψ_2	=	−1/3	−1/3	−1/3	+1
ψ_3	=	0	−1	+1	0

Their orthogonality can be assessed by determining pairwise whether the sums of the products of corresponding coefficients are zero. Orthogonality of contrasts is an important property, as it guarantees that the sums of squares associated with them add up to the total *SS* of the corresponding factor, and that the statistical inferences are based on independent sources of information.

Apart from the distinction between orthogonal and nonorthogonal contrasts, contrasts can also be distinguished with respect to a number of other features, all corresponding to specific research goals. Well-known contrasts are 'deviation', 'Helmert', 'simple', and 'polynomial' contrasts; the last can be used to detect trends across the levels of a factor. Contrasts are very useful in measuring developmental trends over time. If one measures the development of language proficiency in a group of subjects over time (a longitudinal design; repeated measures), one want to test whether there are differences between the points of time used in the design by analysis of variance. If an increase in the proficiency measure shows up over time, one is left with the question of how this increase over time should be characterized. Is it a linear increase, or does the increase accelerate or slow down? The pattern of increase can be investigated by applying specific contrasts to testing linear, quadratic and cubic trends in the pattern over time for instance. More information on trend analysis can be found in Winer et al. (1991) and Kirk (1995). Most manuals of statistical software provide information on how to use and specify contrasts.

The well-known t statistic can be used to test the H_0 that a contrast ψ is zero:

$$t = \frac{c_1\bar{X}_1 + c_2\bar{X}_2 + \cdots + c_k\bar{X}_k}{\sqrt{MS_{\text{error}}(c_1^2/n_1 + c_2^2/n_2 + \cdots + c_k^2/n_k)}} \tag{6}$$

with $df = N - k$. An equivalent test can be carried out by using the F statistic: Numerator and denominator of the preceding formula are squared, $df_1 = 1$, $df_2 = N - k$.

As an illustration we test the hypothesis that $(\mu_1 + \mu_2)/2 = (\mu_3 + \mu_4)/2$ for the data of Table 6.7. The coefficients of the contrast are $+\frac{1}{2}, +\frac{1}{2}, -\frac{1}{2}$ and $-\frac{1}{2}$:

$$t = \frac{(\frac{1}{2} \times 4.50 + \frac{1}{2} \times 6.50 - \frac{1}{2} \times 5.75 - \frac{1}{2} \times 2.0)}{\sqrt{1.229 \times (0.25/4 + 0.25/4 + 0.25/4 + 0.25/4)}} = 2.931$$

with $df = 16 - 4 = 12$. The t ratio is significant at the 0.014 level. The corresponding SPSS output is given in Table 6.8. The easiest way to define ones own contrasts in SPSS is to take the option CONTRASTS in the ONEWAY procedure under COMPARE MEANS. The command in syntax mode is:

```
ONEWAY
  dep BY area
  /CONTRAST= .5 .5 -.5 -.5
  /MISSING ANALYSIS .
```

Table 6.8. Output of contrast analysis for the data in Table 6.7

Contrast coefficients

	Urban area			
Contrast	1.00	2.00	3.00	4.00
1 ①	.5	.5	-.5	-.5

Contrast Tests

		contrast	Value of Error	Std. t	df	(2-tailed)	Sig.
SCORE ②	Equal variances assumed	1	1.6250	.5543	2.331	12	.013
	Equal variances not assumed	1	1.6250	.5543	2.931	10.602	.014

Table 6.8 provides the following information:

① Contrast coefficients: We used the coefficients 0.5 0.5 -0.5 -0.5 in order to compare the difference between Areas 1 and 2 versus Areas 3 and 4.

② As Levene's test for homogeneity of variances was not significant ($F_{3.12}$ = 0.786, p = 0.525, see Chapter 7), we can use the t value mentioned in the line 'Equal variances assumed': It is significant at the 5% level (two-tailed).

6.5. Effect size and strength of association

It is tempting to compare the magnitude of effects by inspecting the associated p values: The smaller the p value is, the 'stronger' the corresponding effect. Let us suppose that the results were found in an analysis of variance as given in Table 6.10. The design in question is a two-factorial design, with factor A representing four groups with different dialects and factor B representing two methods of improving reading skills; the data are fictitious. For clarity's sake we restricted the number of subjects to three per cell. Factors A and B are crossed. The data are given in Table 6.9.

Table 6.9. Two-factor design with A (group, four levels) and B (method, two levels)

	A_1	A_2	A_3	A_4
	4	3	1	5
B_1	8	2	1	7
	7	1	2	5
	3	3	3	4
B_2	4	4	4	7
	5	6	1	9

Table 6.10. SPSS output of an analysis of variance with two crossed factors, A (four groups) and B (two methods)

Source	Type III Sum of Squares	df	Mean Square	F	Sig.
Corrected Model	84.625[a]	7	12.089	5.090	.003
Intercept	408.375	1	408.375	171.947	.000
A	40.458	3	13.486	5.678	.008
B	26.042	1	26.042	10.965	.004
A*B	18.125	3	6.042	2.544	.093
Error	38.000	16	2.375		
Total	531.000	24			
Corrected Total	122.625	23			

a. R Squared = .690 (Adjusted R Squared = .555)

Looking at the p values in Table 6.10, effect B seems to be the strongest effect, followed by A and AB; this interpretation is based on the assumption that smaller p values are associated with stronger effects. Such an interpretation of p values is not correct. The p value of the F statistic is a function of a number of elements:

$$p(F) = f\,[\,(1)\text{ magnitude of effect, } (2)\text{ magnitude of error, } (3)\; df_1,\ (4)\; df_2\,] \qquad (7)$$

Because of the dependency of p on a number of elements, the p value found is not a direct reflection of the strength or magnitude of an effect. In our example the probability values of A and B are determined by:

Factor A	Factor B
1. sum of squares A	1. sum of squares B
2. sum of squares Error	2. sum of squares Error
3. df_1	3. df_2
4. df_4	4. df_4

Lower p values are obtained with larger sample sizes and, consequently, higher degrees of freedom. In our example the dfs associated with factors A and B differ, so the p values cannot be compared. One way of measuring the strength of an effect is to relate the sum of squares ascribed to a factor to the adjusted total sum of squares. A descriptive index for this purpose is η^2 (*eta squared*), often called the correlation ratio. The definition of η^2 is:

$$\eta^2 = \frac{SS_{\text{factor}}}{SS_{\text{corrected total}}} \qquad (8)$$

This descriptive index is related to the concept of explained variance. For factor A this index returns the value (40.458/(40.458 + 26.042 + 18.125 + 38.000) = 0.330, and for factor B (26.042/(40.458 + 26.042 + 18.125 + 38.000) = 0.212. Thus the strength of A exceeds the strength of B, in spite of the 'higher significance' of the latter (p is 0.004 for B, compared to 0.008 for A).

The default index for effect size given in SPSS is not η^2, but *partial* η^2.

$$\text{partial } \eta^2 = \frac{SS_{\text{factor}}}{SS_{\text{factor}} + SS_{\text{error}}} \qquad (9)$$

In our example, this index returns the value 40.458/(40.458+38.000) = 0.516 for factor A, and the value 26.042/(26.042+38.000) = 0.407 for factor B. Again the effect size of A exceeds the strength of B.

Hays (1981: 290-291) presented another index for the strength of association, called ω^2. This index is based on variance components. It is defined as:

$$\omega^2 = \frac{\sigma^2_{\text{factor}}}{\sum \sigma_i^2} \qquad (10)$$

σ^2_{factor} is the estimate of the variance component of a specific effect, and $\sum \sigma_i^2$ is the sum of all variance components specified in the model. As a matter of fact this index can only be used if estimates of genuine variance components are available, that is to say if *random* factors are included in the design. That is why SPSS only presents variance components when the associated factors are *random*. If a factor is random we can estimate effect variances on the basis of the expected values of the Mean Squares of the sources of variance, as was shown in section 4 of Chapter 5. For a design with both A and B as random factors, the following expected values are obtained for the data given in Table 6.9 on the basis of the expected values given in Table 5.5 and the ANOVA given in Table 6.10 with $n = 3$, $p = 4$ and $q = 2$: $E(MS_A) = 6\sigma^2_\alpha + 3\sigma^2_{\alpha\beta} + \sigma^2_\varepsilon$. The expected value of $E(MS_{AB})$ is $3\sigma^2_{\alpha\beta} + \sigma^2_\varepsilon$.

By subtracting the values of *MS* and the corresponding Expected Values we can estimate the variance components. In our example the effect variance σ^2_α is estimated as follows: Subtract MS_{AB} from MS_A and subtract the associated Expected Values: $MS_A - MS_{AB} = (6\sigma^2_\alpha + 3\sigma^2_{\alpha\beta} + \sigma^2_\varepsilon) - (3\sigma^2_{\alpha\beta} + \sigma^2_\varepsilon) = 6\sigma^2_\alpha$. Thus one obtains: $6\sigma^2_\alpha = (13.486\text{-}6.042) = 7.426$. We get the estimate of σ^2_α by dividing the latter value by 6: 1.238. The sum of these variance components is 1.238 + 1.670 + 1.222 + 2.375 = 6.505. Thus $\omega^2_A = 1.238/6.505 = 0.190$. We invite the reader to calculate the other values of ω^2.

In Table 6.11 we give the values of F, p, partial η^2, η^2 and ω^2 for the three sources of variance at issue. Both Factors A and B are considered to be *random* factors.

Table 6.11. Variance estimates given by SPSS in case A and B are random factors

Source	F	p	partial η^2	η^2	ω^2
A	5.678	0.008	0.516	0.330	0.190
B	10.965	0.004	0.407	0.212	0.257
A × B	2.544	0.188	0.323	0.148	0.165

The orders of the effect sizes exhibited by partial η^2 and η^2 are equivalent; unlike the values of η^2 for effects A and B the values of ω^2 for these factors are quite similar.

The conclusion of this section must be that a p value or the value of the F statistic does not say anything about the strength of association. They are primarily related to the hypothesis that there is no particular main or interaction effect.

6.6. Reporting analysis of variance

Often scientific journals have their own standards for statistical reporting. Many journals, however, adhere to the standards of the American Psychological Association (APA). Generally speaking we can say that sections in which analysis of variance is reported should contain the following components:

- The main characteristics of the design (one-way design, completely randomized factorial design, repeated measures design, split plot design etc.).
- which factors were considered fixed and which random and why (cf. Chapters 4 and 5);
- which factors were within-subject and which between-subject factors.
- the percentage of missing data, and which procedure was used to deal with the missing data (cf. Chapter 9, Section 8).
- results of a-priori tests used to assess whether the assumptions of analysis of variance were violated or not (Levene's test of homogeneity of variance, see Chapter 7; the sphericity assumption for repeated measures designs, see Chapter 8).
- the F ratio and the associated degrees of freedom (for instance: $F(3,40) = 12.000$ or $F_{3.40} = 12.000$).
- effect sizes; the statistic generated by SPSS is η^2_{partial} (see Section 5.5), but APA requires the value of *MSE*. *MSE* is a useful statistic. On the basis of the F ratio and the associated degrees of freedom, the values of SS_{factor} and SS_{error} can be calculated and thus the value of η^2. Example: Assume $F(3,40) = 12.000$, and $MSE = 5.000$. $SSE = 40 \times 5.000 = 200$. The F ratio $= 12 = MS_{\text{factor}}/MSE$; consequently $MS_{\text{factor}} = 60$. On the basis of the 3 degrees of freedom, $SS_{\text{factor}} = 180$. Thus the ingredients of η^2 become available: $SS_{\text{factor}}/(SS_{\text{factor}} + SS_{\text{error}})$, in our case $180/(180+200) = 0.474$. Reported values of *MSE* can also be used in calculations to determine the sample size needed to detect an effect.

6.7. Suggestions for reporting statistics

EXAMPLE 1

An analysis of variance carried out on the data obtained in our two-factorial design (fixed factor A with three levels and fixed factor B with two levels) showed a significant effect (significance level 5 per cent) of the interaction A×B: $F_{2.110}$ is 5.980, $p < 0.05$; $MSE = 320.44$ and $\eta^2_{\text{partial}} = 0.30$. In order to assess whether factor B was significant at level 2 of factor A, a simple main effect analysis was carried out. The associated $F_{2.110} = 3.456$ was significant at the 5 per cent level. This means that there is a significant effect of factor B at level 2 of factor A.

EXAMPLE 2

An analysis of variance, carried out on the data obtained in the two-factorial design (fixed factor A with two levels and fixed factor B with two levels) showed a significant effect for factor A ($F_{1.60}$ is 5.980, $p < 0.05$; $MSE = 107.44$ and $\eta^2_{\text{partial}} = 0.36$. The effect for factor B was also significant ($F_{1.60}$ is 3.76, $p < 0.05$; $MSE = 77.36$, and $\eta^2_{\text{partial}} = 0.18$; the interaction A×B was not significant ($F < 1$).

6.8. Terms and concepts

- *Contrast*: A contrast between *k* means is a linear combination of weighted means, in which the weights add up to zero. A set of contrasts is orthogonal if the sum of the products of corresponding coefficients is zero.
- *Effect size in analysis of variance*: A way to assess to which extent a factor or an interaction affects the *SS* or the sum of variances.
- *Post hoc comparison*: A statistical procedure used to evaluate differences between means following a significant overall test.
- *Simple main effect analysis*: A procedure to test whether an effect is significant at a specific level of another factor.

6.9. Exercises

1. Suppose we have a two-factorial design where both factors, A and B, are fixed. We have the following data set with three observations per cell:

	B_1	B_2
A_1	3, 2, 6	7, 4, 9
A_2	2,3,7	8,9,10

Determine η^2 and η^2_{partial} for the factors A and B and the interaction A by B.

2. Determine for the data in exercise 1 the simple main effects for factor B at the two levels of factor A.

3. Is the contrast $(A_2 + A_3)$ vs. A_1 significant at the 5% level in the following data set?

A_1	A_2	A_3
3	7	8
2	4	5
6	9	10
2	8	8
3	9	10
7	10	11

4. What is the difference between ω^2 and η^2?

5. How many orthogonal contrasts can be carried out on the 5 levels of a factor?

6. Below a section of the SPSS output from an analysis of variance is given, carried out on the vowel durations of speakers from three dialects (factor 'dialect'). On the basis of perceived speech tempo ('rate' factor) two groups of speakers were distinguished: fast speakers and slow speakers. The total number of speakers was 24 (= N). Determine a) R^2 and Adjusted R^2 (see Chapter 5), b) η^2 for the two factors and their interaction.

Source	Type III Sum of Squares	df	Mean Square	F	Sig.
Corrected Model	627.000[a]	5	125.400	4.688	.006
Intercept	143221.500	1	143221.500	5354.075	.000
RATE	322.667	1	322.667	12.062	.003
DIALECT	84.250	2	42.125	1.575	.234
RATE*DIALECT	220.083	2	110.042	4.114	.034
Error	481.500	18	26.750		
Total	144330.000	24			
Corrected Total	1108.500	23			

Chapter 7
Violations of assumptions in factorial designs and unbalanced designs

7.1. Preview

Most statistical tests are based on a number of *assumptions*. These can pertain to the measurement scale (nominal, ordinal, interval, ratio), the method of sampling and/or assigning subjects to treatments (randomly or nonrandomly), the selection of factor levels, the distribution of the populations from which the samples are drawn, the distribution of the error, etc.

The assumptions underlying analysis of variance were mentioned in the preceding chapters. Of course these assumptions do not apply automatically; there are situations in which they are violated. Such violations can have serious consequences for the validity of the statistical conclusions. For instance, one may think that the assumption of equal variances is fulfilled, whereas in reality it is not. The consequence may be that the probability of committing a type I error in carrying out an analysis of variance is not .05 (the 'nominal' level), but 0.12.

In Section 7.2 we start with an overview of the *assumptions underlying analysis of variance*. Section 7.3 is devoted to the assumption of *homogeneity of variance*. In Section 7.4 we discuss how *transformations* of the dependent variable may help in meeting some assumptions underlying analysis of variance.

In Section 7.5 the effects of *deviations from the interval scale level* on the results of analysis of variance are discussed. In Section 7.6 the problem of *unequal cell sizes* ('missing data') is dealt with; this problem deserves special attention because researchers do not always seem to be fully aware of the consequences of having unequal cells.

7.2. Assumptions in analysis of variance

There are three general assumptions on which analysis of variance is based:

1. The normality assumption: The observations are drawn from normally distributed populations.

2. The homogeneity of variance assumption: The variances of all (sub)-populations are equal.
3. The independence assumption: The observations are independent.

What can be done if any of these assumptions is violated or, rather, should anything be done? In general, analysis of variance is said to be robust with regard to violations of its assumptions. The qualification 'robust' means that the nominal α level (the level assumed by the researcher) is very close to the actual α level (the level which holds in a concrete application). Nonnormality hardly appears to affect the α level. The F statistic is quite robust with respect to heterogeneous variances too, provided the cell frequencies are large and equal.

In terms of robustness, the independence assumption is by far the most important one (cf. Stevens 1986: 202 ff., Kenny and Judd 1986). Even small violations can have dramatic effects. When groups are sampled, the individual observations may turn out to be dependent because participants influence or affect other participants within the same group with respect to the variable on which they are measured. This type of dependence is called 'nonindependence due to groups' by Kenny and Judd (1986). This occurs when participants of specific groups have something in common, for instance the textbook they have been using, their teacher, and even the specific social interaction they have been exposed to. Stevens (1986: 204-205) suggests taking the group mean as the unit of analysis when within-group dependencies occur. Another type of dependence is 'nonindependence due to sequence'. An example of this is a specific pattern arising as a result of observations taken repeatedly from a single experimental unit. Such a pattern arises when adjacent observations are linked. Notably in rating experiments we can find this type of nonindependence, for instance when participants are asked to rate series of synthesized utterances or samples of pathological speech. It is conceivable that they try to vary their ratings in order to create scores which are different in appearance. This could create a sequence where low ratings are followed by relatively high ones.

Dependence between observations should be avoided in collecting data, because it introduces a strong bias in the statistical interpretation of the results. That is why sampling methods in which previously selected informants are used to find new potential informants should not be applied in language behaviour studies.

On the other hand, there are many situations in which dependence is an essential part of the phenomenon we want to investigate. An example is the study of interactional patterns between persons. Dependence here means that people in interaction influence each others' verbal behaviour. When interruptive verbal behaviour is investigated in opposite-gender dyads, one may think that the role of the factor gender has to be analyzed by comparing the scores of the male and female groups; consequently, a straightforward one-way analysis of variance for independent groups is applied. Beware, this is the wrong approach. The basic units are not separate subjects but dyads of subjects, each dyad consisting of a male and a female subject. The variable gender, represented within each dyad, has to be treated like a repeated measure (cf. repeated measures analysis of variance in Chapter 8). A discussion of design issues in dyadic research can be found in Kenny (1990). If possible, dependence relationships in the data should be incorporated as a source of variation in the analysis of variance. On many occasions it is defensible that a researcher treats violations of statistical assumptions as a harmless nuisance. One should, however, always ask oneself what kind of properties in the data studied can cause violations of underlying assumptions and try to answer the question whether the violations found are inherent to the phenomenon under investigation. One should query whether these properties can be accounted for in the model which underlies the data. This implies that one should always inspect the data thoroughly for possible trouble sources, however negligible they may seem in testing specific hypotheses.

7.3. Normality of variances and homogeneity

The first assumption mentioned in Section 7.2 is the assumption of normality. This refers to the form of the distribution of the populations from which the samples are drawn. In fact there are two ways in which the assumption may not be met: a) The distribution is skewed, b) The dependent variable is not an interval variable, or is an interval variable with just two values, i.e. '0-1' data. Violations of the assumption that the distributions are not skewed do not appear to have serious consequences, neither in t tests nor in analysis of variance. For 0-1 data we refer to Lunney (1970). He showed that in specific conditions significance levels obtained in analysis of variance with 0-1 data approach the nominal levels. He mentioned the following conditions:

- The factors of the design are fixed.

- Cell sizes are equal.
- The proportion of responses in the smaller response category (being either 0 or 1) exceeds 0.20, and the *df* of the within-cell variance exceeds 20, OR the proportion of responses in the smaller response category is lower than 0.20, but the *df* of the within-cell variance exceeds 40.

Designs with repeated measures were not an object of Lunney's study. In general, 0-1 data are analyzed with loglinear models or logistic regression, techniques currently used in the analysis of frequency data.

The second assumption mentioned in Section 7.2 is the homogeneity of variance. If an analysis of variance is carried out, this so-called homoscedasticity assumption should be tested routinely. The H_0 that has to be tested is the equality of variance in each (sub)population: $\sigma_1^2 = \sigma_2^2 = \ldots = \sigma_k^2$. If the variances are not equal, the value of SS_{error} also depends on $SS_{\text{treatments}}$ / SS_{groups}. Thus $SS_{\text{treatments}}$ / SS_{groups} and SS_{error} are not independent estimators of σ_ε^2. Levene's test is designed to test this assumption. This test involves an analysis of variance of the scores of the dependent variable after the following transformation:

$$Y_{ij} = |X_{ij} - \bar{X}_i| \qquad (1)$$

where $\bar{X}_i$ is the mean of the scores in group *i*. If the resulting *F* ratio is significant, the hypothesis of homogeneity of variance has to be rejected. If the data have a highly skewed distribution, $\bar{X}_i$ should be replaced by the median of group *i*.

We illustrate how to process the data given in Table 7.1, a one-way design with three groups and five observations per group.

Table 7.1. Scores obtained in three treatments with five observations each

	A_1	A_2	A_3
	4	2	3
	5	3	7
	7	4	6
	10	3	2
	16	2	8
$\bar{X}$	8.40	2.80	5.20
s^2	23.30	0.70	6.70

In Table 7.2 we reproduce the SPSS output of a test of homogeneity of variance carried out on the data of Table 7.1.

Table 7.2. Levene's test in SPSS output obtained by the One-Way ANOVA procedure, applied to the data given in Table 7.1

Levene Statistic	df1	df2	Sig.
4.666	2	12	.032

Table 7.2 shows that the $F_{2,12}$ associated with Levene's test is 4.666, which is significant at the 5% level; one has to reject the hypothesis that the variances across the groups are equal. Many textbooks tell us not to worry about possible violations of the homoscedasticity assumption. A study by Wilcox (1987), however, made clear that violations of this kind should be taken seriously, especially if marked differences in variance arise. Wilcox showed that unequal variances can substantially affect type I error probabilities. The actual α levels turn out to be higher than the nominal ones. In order to illustrate this effect, we reproduce part of a table taken from his article in Table 7.3. The entries are: number of treatments in the one-way analysis of variance (= k), sample sizes (n_j), relative magnitudes of standard deviations (s_j) and the nominal α levels.

Table 7.3. The effects of heterogeneous variances on the α level (after Wilcox 1987)

k	sample sizes $n_j, j = 1...k$	stand. dev. $s_j, j = 1...k$	nominal α level .100	.050	.010
3	6, 6, 6	1, 2, 3	.116	.066	.018
"	"	1, 1, 3	.137	.082	.029
4	12, 12, 12, 12	1, 1, 1, 4	.144	.101	.046
"	"	1, 1, 1, 3	.132	.089	.036

Let us suppose that a one-way analysis of variance is carried out on four treatments, with 12 subjects in each treatment. The relative magnitudes of the standard deviations within the treatments are 1, 1, 1, 3. In Table 7.3 we can see that in this situation a nominal α level of 0.05 turns out to be an actual α level of 0.089. Most researchers would not accept this 'significance level'.

If the variances in the cells of an analysis of variance are not equal – as confirmed by Levene's test – a more robust test for oneway analysis of variance is available, developed by Welch (1947). Welch's test is approximately

F distributed and consists in a) the introduction of weighted variances (W_i) and b) adjusted degrees of freedom: $df_{2\ \mathrm{adj}}$. In formulas (2), (3) and (4) we give three ingredients for the Welch adjusted F ratio (formula (5)) and the associated degrees of freedom (formula (6)).

$$W_j = \frac{n_j}{s_j^2} \quad (2)$$

$$\bar{X}^* = \frac{\Sigma W_j \bar{X}_j}{\Sigma W_j} \quad (3)$$

$$\lambda = \Sigma \frac{(1 - W_j / \Sigma W_j)^2}{n_j - 1} \quad (4)$$

$$F_W = \frac{\Sigma W_j \frac{(\bar{X}_j - \bar{X}^*)^2}{(k-1)}}{1 + 2(k-2)\lambda/(k^2-1)} \quad (5)$$

$$df_{2\ \mathrm{adj}} = (k^2 - 1)/3\lambda \quad (6)$$

For the example of Table 7.1 we obtain the following components: $k = 3$, $n_j = 5$, $\Sigma W_j = 8.094$, $\bar{X}^* = 3.170$, $\lambda = 0.447$. Filling in these components in (5) and (6), we obtain $F = 4.463$, $df_1 = 2$, and $df_2 = 5.972$. The associated p value is 0.065, which is not significant at the $\alpha = 0.05$ level. The SPSS output is given in table 7.4.

Table 7.4. Output of SPSS One-Way ANOVA procedure, options Welch, applied to the data given in Table 7.1

Statistic	df1	df2	Sig.
4.463	2	5.972	.065

Levene's test is a good instrument to assess the extent to which the data meet the assumption of equal variances. Its use is not restricted to one-way designs, it can also be used in multifactorial designs.

A relevant question now is: What happens if assumptions on parameters of the populations from which samples are drawn are not met. In fact there are three options:

- Carry out a *transformation* on the data; well-known transformations are the log and arcsine transformations to obtain equal variances in the cells of the design. We will discuss these transformations in Section 7.4.

- Use one of the well-known *non-parametric* alternatives to analysis of variance. We mention the Kruskal-Wallis one-way analysis of variance (for independent samples) and the Friedman two-way analysis of variance (including repeated measures, see Chapter 8 for repeated measures). Both tests are based on a transformation of the dependent variable into ranks. A clear description of these techniques can be found in the classic textbook by Siegel and Castellan (1988). That is why we do not discuss them here, despite their usefulness.
- Use a *randomization test*. The randomization test is a recent development in statistics, aiming at making tests independent of population parameters. We will present this test in Chapter 10.

7.4. The impact of transformations

A large number of transformations can be applied to make the data meet the assumptions of the analysis at hand. Which of these must be applied depends on the objective of the transformation and the nature of the data. Generally speaking it only makes sense to use transformations which are not linear. Nonlinear transformations stretch or contract the original data in a nonuniform way. This effect can be illustrated by a logarithmic transformation: $\log X$. Let us assume that we collected the scores, obtained in three treatments, given in the left-hand part of Table 7.5. The data are the same as the ones shown in Table 7.1.

Table 7.5. Scores obtained in three treatments with five observations each as in Table 7.1, and the same set of scores after logarithmic transformation

	A_1	A_2	A_3	log →	A_1	A_2	A_3
	4	2	3		0.60	0.30	0.48
	5	3	7		0.70	0.48	0.85
	7	4	6		0.85	0.60	0.78
	10	3	2		1.00	0.48	0.30
	16	2	8		1.20	0.30	0.90
$\bar{X}$	8.40	2.80	5.20		0.870	0.432	0.661
s^2	23.30	0.70	6.70		0.058	0.017	0.067

The variances of A_1, A_2 and A_3 clearly differ. Moreover, there seems to be a relationship between the magnitude of the scores and the amount of variance. A logarithmic transformation can be applied to make the variances more similar. The logarithmic transformation (log, which is the logarithm with the base 10) of the original scores returns the results given in Table 7.5 on the right.

The transformed scores in Table 7.5 illustrate the contracting effect of the logarithmic transformation; the higher the values of the scores are, the more prominent this effect becomes (provided that the scores involved are larger than 1). This is illustrated in Figure 7.1. It shows that the values between 1 and 10 range between 0 and 1 after transformation, whereas the values between 10 and 100 are contracted within the value range of 1 and 2. The consequence of this operation is a greater decrease in the variance of groups with high scores than in the variance of groups with low ones. If scores also include the value 0 ($\log 0$ is not defined) a constant should be added to all scores, e.g. the constant 0.5 or 1: $\log(1+X)$.

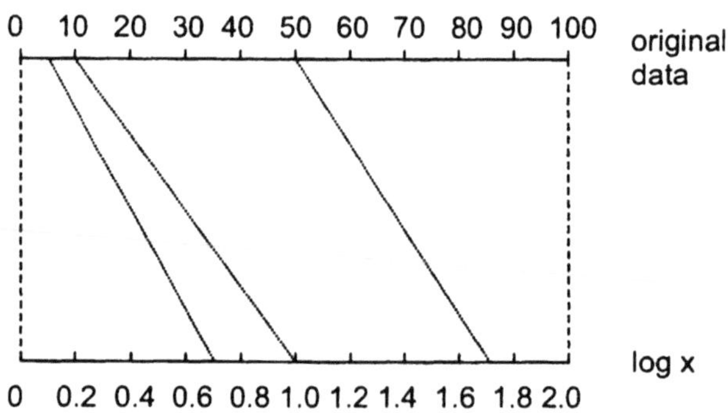

Figure 7.1. Effect of the log transformation

To illustrate the effect of the logarithmic transformation on the results of an analysis of variance, the output of the SPSS One-Way procedure is reproduced. It was successively carried out on the original scores and the transformed scores of Table 7.5. The ANOVA results for the original scores are presented in Table 7.6, together with Levene's test for homogeneity of variances.

Table 7.7 gives the results of the one-way analysis of variance of the log-transformed scores of Table 7.5. Table 7.7 shows that after the logarithmic transformation Levene's test is not significant anymore. Moreover the F ratio has become significant at the 5% level ($F_{2,12} = 5.087$, $p = 0.025$). This shift is not necessarily a consequence of the log transformation. The result of the log

Table 7.6. (Part of the) SPSS output from an analysis of variance (ONEWAY) applied to the original scores of Table 7.5

Levene Statistic	df1	df2	Sig.
4.666	2	12	.032

	Sum of Squares	df	Mean Square	F	Sig.
Between Groups	78.9333	2	39.467	3.8567	.051
Within Groups	122.8000	12	10.2333		
total	201.7333	14			

Table 7.7. (Part of the) SPSS output from an analysis of variance (ONEWAY) applied to the transformed scores of Table 7.5

Levene Statistic	df1	df2	Sig.
1.960	2	12	.183

	Sum of Squares	df	Mean Square	F	Sig.
Between Groups	.481	2	.240	5.087	.025
Within Groups	.567	12	.047		
total	1.048	14			

transformation depends on the distributional characteristics of the analyzed data. In our example the tests for homogeneity of variance are not significant after transformation, so we may assume that the variances are equal and that the use of the F ratio is justified.

Another well-known transformation, especially used with proportions, is the arcsine transformation. The arcsine x is also called the inverse sine: $\sin^{-1} x$, a function which can be found on many pocket calculators and also in SPSS. The transformation applied is: $X = 2\arcsin\sqrt{p}$, in which p is a proportion. Let us give an example of the arcsine. You may remember that $\sin 45^o$ is 0.707. Instead of degrees, radians can be used, that is to say that the range of 0 to 360 degrees is translated into the range of 0π to 2π. Thus, $\sin \frac{1}{4}\pi$ is also 0.707. The inverse operation is: $\arcsin 0.707 = \frac{1}{4}\pi$. For small data sets, however, the following formula is recommended by Studebaker (1985):

$$X = \arcsin\sqrt{S/(N+1)} + \arcsin\sqrt{(S+1)/(N+1)} \qquad (7)$$

S is the number of 'successes' which, if divided by N (the number of observations), yields the proportion p.

Why transform proportions. In fact there are two reasons:

- The variances of distributions underlying p are not constant, but vary with p ($s_p^2 = p(1-p)$). This makes the variances dependent on the treatment means.
- The unit of measurement of proportions is not constant over the scale. As Cohen and Cohen (1985) note: "It is (...) evident (...) that the 0.04 difference between 0.01 and 0.05 is much more important than the same difference between 0.48 and 0.52." The former represents a fivefold increase, the latter an increase of 8%. Therefore, a transformation that stretches the tails of the range of numerical values is appropriate.

In Table 7.8 we present the scores of a data set with three treatments. The data are proportions representing, for instance, 'correct scores'. The variances differ considerably. After the arcsine transformation of the scores in the left-hand part of Table 7.8, we get the scores on the right.

Table 7.8. Proportions (left) and scores after arcsine transformation (right); transformation applied is $2 \arcsin \sqrt{p}$

	A_1	A_2	A_3	$2 \arcsin \sqrt{p} \rightarrow$	A_1	A_2	A_3
	0.04	0.18	0.35		0.40	0.88	1.27
	0.11	0.40	0.83		0.68	1.37	2.29
	0.16	0.25	0.41		0.82	1.05	1.39
	0.09	0.32	0.63		0.61	1.20	1.83
$\bar{X}$	0.10	0.29	0.56		0.63	1.12	1.70
s^2	0.003	0.009	0.048		0.030	0.045	0.217

The arcsine transformation is marginally successful: For the original data Levene's $F = 6.131$ ($p = 0$ .019), for the transformed data Levene's $F = 3.607$ ($p = 0.071$).

The arcsine transformed data range between approximately 0.10 and 3.04.

An alternative preferable to an arcsine transformation of proportions is the use of the so-called logit. We must emphasize the usefulness of the logit here, as in many language behaviour studies proportions constitute the dependent variable.

When a transformation is applied, the statistical conclusions are only valid for the transformed scores. Consequently, it only makes sense to carry out a transformation if one can give an acceptable interpretation of the transformed scores and present plausible arguments as to how the transformation is related to underlying processes. Many time-related measures like growth, age, learning and reaction time, but also frequency values (expressed in Hz) which reflect fundamental frequency appear to fit better after a logarithmic transformation.

The presentation of different methods for data transformation may suggest that dissimilar variances in treatment groups only constitute a computational problem, which can be 'smoothed away' by applying the necessary transformations. Bryk and Raudenbush (1988), however, showed that heterogeneity of variance may be taken as evidence of an interaction between the 'treatment' factor and the subjects involved. A specific treatment may lead to high scores in participants with specific characteristics, whereas participants with other characteristics obtain low scores. In different situations ceiling effects may occur; the scores are high and almost equal, and the associated variance is low. Not only does a purely statistical problem crop up in these cases, but also one that is closely related to the design of the experiment. That is why significant tests of homogeneity of variance should persuade the investigator to take a closer look at her/his data, instead of routinely carrying out a data transformation.

7.5. Scale of measurement and analysis of variance

Since Siegel's book 'Nonparametric Statistics' (1956) many social scientists and researchers who tend to emulate methods in social sciences (viz. sociolinguists and speech- and language pathologists), refrain from using so-called interval statistics like *t* tests and *F* ratios in analysis of variance when they suspect that the data at issue are not measured at the interval level. They think that the statistical result of a *t* test is affected by the mechanism which produced the numbers. The rationale for this approach is that interval statistics, like the *t* test and *F* ratio use the same properties of the numbers used, as the latter are assumed to reflect the properties of the objects: '6-7' = '2-3'. Many statisticians and methodologists do not follow this strict line. They say that "statistics does not know what is behind the numbers", and prefer to distinguish between numbers seen as scores which are assumed to reflect magnitudes of properties of objects, and numbers seen as input for a statistical procedure which is designed to find out whether the differences in the numbers obtained from two (or more) groups are due to chance or not. Harris (1975: 226-227) for instance, said: "The validity of statistical conclusions depends only on whether the numbers to which they are applied meet the distributional assumptions to derive them, and not the scaling procedures used to obtain the numbers." This author also drew attention to the fact that most scale values which are assumed to be of the ordinal level - at least by some strict interpreters of Stevens' laws (Stevens 1951) - contain interval information. An example: Assume we obtained the values 3 - 4 - 7 - 9 on a scale with 10 categories. If these values are ordinal data, any monotonic transformation is permissible, which means that a transformation resulting in the values 3 - 4 - 7 - 999 is equally valid. Obviously this is nonsense, and apparently the intervals between the numbers used contain some information. This led both Harris (1975) and Labovitz (1970: 515) to quite categorical statements. We quote the latter: "Empirical evidence supports the treatment of ordinal variables as if they conform to interval variables (.....). Although some small error may accompany the treatments of ordinal variables as interval, this is offset by the use of more powerful, more sensitive, better developed, and more clearly interpretable statistics with a known sampling error." On the basis of these statements - and a lot more similar statements can be quoted, see for instance Anderson (1961) - we think that analysis of variance can be applied to the data which are not strictly of the interval level, but which contain information on the magnitude of the differences between objects, as obtained in magnitude scaling and equal appearing interval scaling (EAI).

In the previous section we made clear that one does has not to worry too much about the scale level of the data which is submitted to analysis of variance. The considerations were of a statistical nature: F ratios are not so much affected by violations of the assumption of the interval measurement level as one often thinks. However, characteristics of the scale or the scale level can affect the interpretation of interaction patterns. A good example of the former was given in Section 5.6, in which we showed that the conversion of the dependent variable 'number of words' into 'mean time per word' caused a dramatical change in the interaction pattern. There are also situations in which the scale level (interval vs. ordinal) can affect the interpretation or even emergence of an interaction effect. An example: Assume that listeners are presented with utterances and have to rate their degrees of perceived 'friendliness'. The pitch register of these utterances is manipulated by computer (cf. Chen, Gussenhoven & Rietveld, 2004). Let us assume that 10 physical levels of pitch register were included, expressed in Hz at equal intervals. When listeners are asked to rate the friendliness of these manipulated utterances (for instance on a 10 point rating scale, often called an 'Equal Appearing Interval scale' (EAI)), we should not be surprised to find that differences at the higher register levels have a smaller perceptual effect than at the lower levels. There is a well-documented equivalent of this effect in psycho-acoustics: The difference between a tone of 1000 and 1005 Hz is perceptually smaller than the difference between 100 and 105 Hz. The perceptual conversion can be modelled with a logarithmic transformation (cf. Nolan 2003).

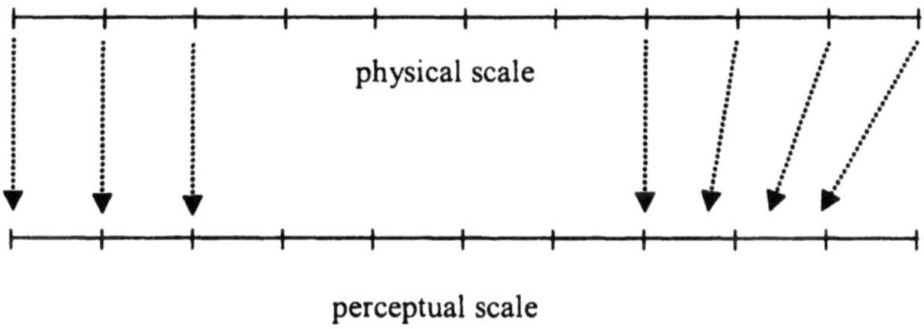

Figure 7.2. Mapping of physical steps (in Hz) on perceptual steps

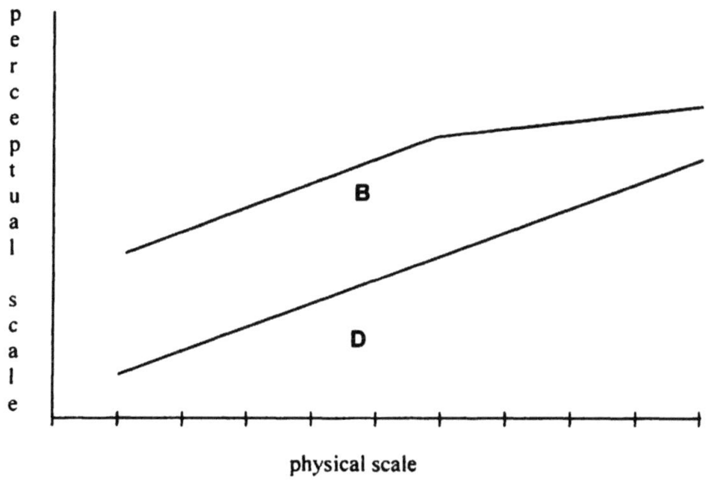

Figure 7.3. The (fictitious) effects of pitch register on perceived friendliness for two groups of listeners, B and D

Figure 7.2 shows the mapping of physical steps (Hz) on perceptual steps (expressed in rating levels) with the latter being smaller at the top end of the scale than at the bottom end.

In Figure 7.3 we show the possible effects of the 'Pitch register' factor (measured in Hz) on perceived friendliness for two groups of listeners (D = Dutch listeners, B = British English listeners). Assume that the British English listeners perceive the utterances as much friendlier than the Dutch listeners. It might be that adding an extra 'quantum' of friendliness needs a larger physical step at higher levels of perceived friendliness than at lower levels. This may explain the steeper slope observed for the B listeners at the lower end of the scale than at the higher end. The effect shown in Figure 7.2, with smaller perceptual differences at higher physical values of pitch register may also account for the 'break' in the ratings of the B listeners. The resulting interaction pattern L(anguage) × Ph(ysical) scale is not easy to interpret. As a matter of fact there are two options:

- B listeners apply another ('linguistic') pitch transformation than D listeners; this means that the perceptual effect of the physical scale is ordinal.
- B listeners use another pitch transformation than D listeners because from the beginning they locate the perceived friendliness at higher scale values; thus from the beginning of the physical scale they need larger physical magnitudes to perceive differences in perceived friendliness.

We do not suggest that the use of scales which are not strictly of the interval level - as is often the case in rating experiments - makes a meaningful interpretation of results impossible. However, one should be careful, as interaction patterns might be the result of a number of effects.

7.6. Unbalanced designs and regression analysis

The default situation described in the preceding chapters was ideal: There were no missing data. All respondents took part in the experiments, nobody was ever distracted and forgot to press a button, no score was lost. We all know that this ideal situation hardly ever occurs. In longitudinal research people die or move house, in psycholinguistic research (with boring experiments, at least from the perspective of subjects) participants get distracted and forget to press buttons, in sociolinguistic research subjects categorically refuse to fill in forms, and sometimes, data just get lost. Missing data are a serious and frequent problem, which can occur in different guises. The first one is that of missing data in one or more cells, the second (and more serious one) is that of one or more missing cells. There are a number of ways to cope with this problem. In some situations researchers prefer to fill in the missing data (this is called *imputation*) on the basis of formal criteria and data which are present, the second one is to use *statistical* techniques to cope with the problem. Let us start to illustrate the problem on the basis of a simple example, a two-factorial completely randomized factorial design. Unequal cell sizes in these designs often create problems. In such *unbalanced* (= nonorthogonal) designs a dependency or correlation may arise between the explanatory factors (= independent variables). A correlation between the explanatory variables leads to problems in the partitioning of the sums of squares of the dependent variable in a unique way. Where two or more factors are correlated, the factors involved explain similar parts of the dependent variable. Consequently, there is no unique procedure to determine which factor influences the dependent variable and to what extent. Take a very simple data set with two factors A and B; there are two observations in the cells A_iB_1 and one in the cells A_iB_2. The data are given in Table 7.9.

Table 7.9. Artificial data set with unequal cell frequencies

	B_1	B_2	$\bar{X}_{i..}$
A_1	1	2	2
	3		
A_2	2	6	4
	4		
$\bar{X}_{.j.}$	2.5	4	

For factor A the mean $\bar{X}_{1..} = (1+3+2)/3 = 2$, and the mean $\bar{X}_{2..} = (2+4+6)/3 = 4$. The difference between these two means may incorrectly be seen as the effect of factor A. However, when we express the data in terms of model equations - disregarding interaction for simplicity's sake - we get the following picture:

$$
\begin{array}{lcl}
X_{111} & = & \mu + \alpha_1 + \beta_1 = 1 \\
X_{112} & = & \mu + \alpha_1 + \beta_1 = 3 \\
X_{121} & = & \mu + \alpha_1 + \beta_2 = 2 \\
X_{211} & = & \mu + \alpha_2 + \beta_1 = 2 \\
X_{212} & = & \mu + \alpha_2 + \beta_2 = 4 \\
X_{221} & = & \mu + \alpha_2 + \beta_2 = 6
\end{array}
$$

By calculating $\bar{X}_{1..} - \bar{X}_{2..}$ - the difference between the means of the two levels of factor A pooled over the levels of factor B - we estimate the difference between α_1 and α_2. However, in our case with missing values we do not obtain the estimate we are looking for:

$$\begin{aligned} \bar{X}_{1..} - \bar{X}_{2..} &= 1/3[(\alpha_1 + \beta_1) + (\alpha_1 + \beta_1) + (\alpha_1 + \beta_2)] \\ &\quad - 1/3[(\alpha_2 + \beta_1) + (\alpha_2 + \beta_1) + (\alpha_2 + \beta_2)] \\ &= (\alpha_1 - \alpha_2) + 1/3(\beta_1 - \beta_2) \end{aligned} \quad (8)$$

This means that we do *not* test the hypothesis at issue, but a different one. A way out is given by *regression analysis*. In regression analysis an equation is set up, with the dependent variable to the left of the equal sign and predictors Y_i with corresponding weighting factors B_i to the right:

$$X' = B_0 + B_1Y_1 + B_2Y_2 + B_3Y_3 \quad (9)$$

The purpose of multiple regression analysis is to assign weights to the predictors Y_i in such a way that the sum of squared differences (*SS*) between the observed values X_i and the values X_i' predicted by the regression equation is minimal. This is illustrated in Figure 7.4, where X is the predictor of Y for a simple regression equation with only one predictor:

$$X' = B_0 + B_1Y_1 \quad (10)$$

The regression line in Figure 7.4 represents the predicted values X', the vertical lines show the deviations from the observed values X. The total Sum of Squares can be partitioned into two parts:

$$SS_{\text{total}} = SS_{\text{regression}} + SS_{\text{residual}} \quad (11)$$

$SS_{\text{regression}}$ is the component of SS_{total} accounted for by the regression equation; the remaining - unaccounted for - Sum of Squares (the deviations between X' and X in Figure 7.4) is called SS_{residual}.

In multiple regression the values of B_i represent the weights attributed to a specific predictor in order to obtain values of X_i', which are as close as possible to the observed values of X_i. Analysis of variance and regression analysis serve the same purpose: Model equations are set up, and statistics have to decide which components of these models should be retained and which should not. As a matter of fact, most computer programs for analysis of variance

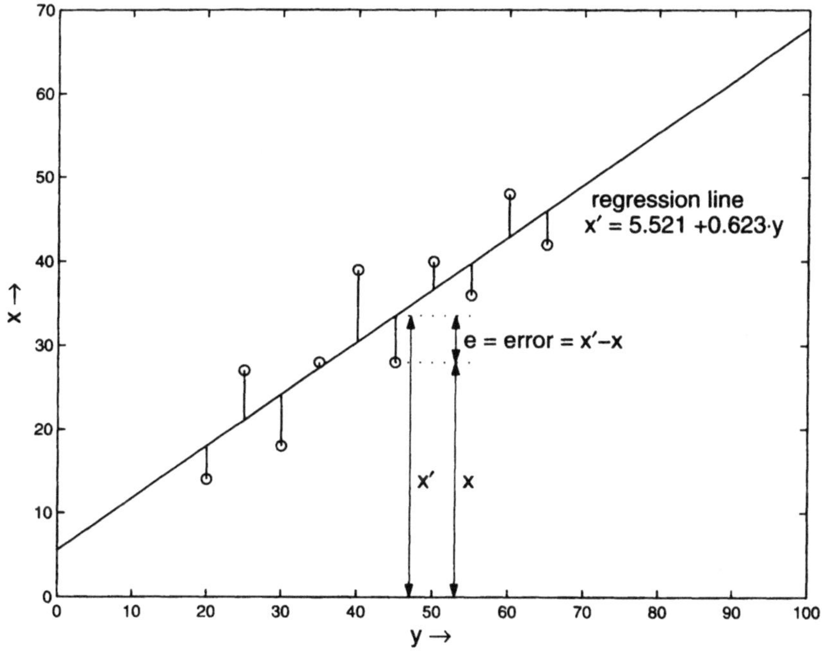

Figure 7.4. A plot of observed values X and predicted values X' on the basis of the regression equation $X' = 5.521 + 0.623Y$

are designed in such a way that they carry out a multiple regression analysis - numerically speaking - with predictor variables which are derived from the ANOVA factor specifications. Regression analysis can be carried out in a number of ways. One can start by assessing whether independent variable Y_1 can be assigned a weight so that the sum of the squared differences between observed values X_i and predicted values X_i' is minimal; in a subsequent step one can continue and assess in how far the remaining part of the differences between X_i and X_i' (SS_{residual}) can be reduced by assigning an appropriate weight to independent variable Y_2. Of course one gets different results if one starts the other way round: First Y_2 is entered and after that Y_1. That is why four types of approaches can be distinguished in analysis of variance, depending on the order in which the independent variables which represent the factors of the design are entered. A short hand notation of the order in which the predictors were entered to calculate the *SS* associated with a specific effect in a completely randomized two-factorial design is given in Table 7.10. The model equation for this design is:

$$X_{ijk} = \mu + \alpha_i + \beta_j + \alpha\beta_{ij} + \varepsilon_{ijk} \qquad (12)$$

In this table we present different approaches (in fact orders of entering independent variables) to obtain Sums of Squares. Four ways are distinguished, referred to as Type I, II, III and IV. Type IV is important when cells are missing completely. For designs with unequal cells Type II=Type IV.

Table 7.10. TYPES OF *SS* Four types of Sums of Squares: Type I, II, III and IV; Type III is Type IV when there are no empty cells

EFFECT	TYPE I	TYPE II	TYPE III=TYPE IV
A	$R(\alpha\|\mu)$	$R(\alpha\|\mu,\beta)$	$R(\alpha\|\mu,\beta,\alpha\beta)$
B	$R(\beta\|\mu,\alpha)$	$R(\beta\|\mu,\alpha)$	$R(\beta\|\mu,\alpha,\ \alpha\beta)$
A × B	$R(\alpha\beta\|\mu,\ \alpha,\beta)$	$R(\alpha\beta\|\mu,\alpha,\beta)$	$R(\alpha\beta\|\mu,\alpha,\beta)$

In this table the different approaches to obtain Sums of Squares are symbolized by $R(\text{ factor}_i)|(\text{ factor}_j)$. This means that the *SS* of factor_i is calculated after having entered factor_j. As an example we look at the expression $R(\beta|\mu,\alpha,\ \alpha\beta)$, for the *SS* associated with factor B, Type III. This expression means that the sums of squares attributed to factor B are obtained after the predictors μ, α, $\alpha\beta$ were entered in the regression equation $X_{ijk} = \mu + \alpha_i + \alpha\beta_{ij} + \varepsilon_{ijk}$. This equation results in an *SS* with a certain value. Entering β in a subsequent step increases the *SS* due to regression by a certain amount: That is the *SS* associated with factor B. At the same time it decreases SS_{residual}. In Table 7.11 we summarize the use of the four types of sums of squares.

Table 7.11. The use of different regression approaches in analysis of variance

Type of *SS*	Use
Type I	The sources of variance are added sequentially: Each term is only adjusted for the term that precedes it in the model, in SPSS syntax it is used for nested models.
Type II	Used for any model that has main effects only.
Type III	Default in SPSS; any design with no empty cells.
Type IV	Used for situations with missing cells.

To illustrate this procedure we present data from the sociolinguistic study of Milroy (1980) on the urban vernacular of Belfast. Central factors investigated by Milroy to explain linguistic variation were 'gender' and 'social network'. She collected informal spontaneous speech from 46 informants. Their speech was analyzed for nine so-called linguistic variables, among them the realization of /a/. This variable measures the degree of retraction and back-raising in items such as 'hat', 'man', 'grass'. The phonetic realization of every occurrence of this variable was scored on a five-point scale, ranging from 1 (approximating the prestige norm) to 5 (approximating the vernacular norm). On the basis of these scores a mean was calculated for every informant, provided the variable occurred in the spontaneous speech material collected. A mean score was calculated for 37 informants. The scores on the linguistic variable can be related to the informant characteristics 'gender' and 'social network' (the two factors). Milroy operationalized the factor 'social network' on the basis of the multiplexity and density of an informant's contacts with other people. A higher score indicates a more open network, a lower score a more closed one. The data analyzed here are given in Table 7.12. The original data can be found in the Appendix in Milroy (1980). In Table 7.12 the network scores, which in fact ranged from 0 to 5, are divided into two levels: low scores (a score of 2 or lower) and high scores (a score of 3 or higher).

Table 7.12. Data from Milroy (1980) on the linguistic variable (a); independent variables are gender and network score (low, high); $N = 37$

	gender							
network	female				male			
low score	1.73	2.13	2.16	2.38	2.65	2.73	2.45	2.85
	1.45	2.63	1.75	1.05	3.05	2.73		
	2.25	2.42	2.78					
	($n = 11$, $\bar{X} = 2.07$)				($n = 6$, $\bar{X} = 2.74$)			
high score	2.33	2.48	2.75	2.70	2.83	3.78	3.13	3.43
	2.35	2.33	2.50		3.15	3.03	2.75	3.03
					2.30	2.55	2.12	2.78
					2.18			
	($n = 7$, $\bar{X} = 2.49$)				($n = 13$, $\bar{X} = 2.85$)			

Table 7.12 contains unequal cell frequencies. These point to a correlation between network and gender: Most female informants have a low network score, whereas most male informants have a high network score. If the data are submitted to the GLM, Univariate procedure in SPSS with *SS* = Type III, the results given in Table 7.13 are returned.

To illustrate the analogies between *regression analysis* and *analysis of*

Table 7.13. Default SPSS output data Milroy (1980), analysis of variance with Type III *SS*

Source	Type III Sum of Squares	df	Mean Square	F	Sig.
Corrected Model	3.985[a]	3	1.328	7.392	.001
Intercept	215.900	1	215.900	1201.391	.000
NETWORK	.594	1	.594	3.305	.078
GENDER	2.250	1	2.250	12.519	.001
NETWORK*GENDER	.211	1	.211	1,176	.286
Error	5.930	33	.180		
Total	247.154	37			
Corrected Total	9.916	36			

a. R Squared = .402 (Adjusted R Squared = .348)

variance in Table 7.14 we present the results of two analyses of variance, one labelled Type I and the other labelled Type III by SPSS. We will show how these results can be obtained by applying regression analyses in different orders. For clarity's sake we only present the associated Sums of Squares.

Table 7.14. Sums of Squares (*SS*) associated with the factors NETWORK, GENDER and their interaction in the study by Milroy (1980) according to Type I and Type III of the procedure GLM in SPSS

Source	Type I *SS*	Model	Type III *SS*	Model
NETWORK	1.619	$R(\alpha\|\mu)$	0.594	$R(\alpha\|\mu,\beta,\alpha\beta)$
GENDER	2.155	$R(\beta\|\mu,\alpha)$	2.250	$R(\beta\|\mu,\alpha,\alpha\beta)$
N × S	0.211	$R(\alpha\beta\|\mu,\alpha\beta)$	0.211	$R(\alpha\beta\|\mu,\alpha,\beta)$

It may have become clear that analysis of variance offers a subset of the analytic possibilities of multiple regression analysis. Both techniques, therefore, share the assumptions mentioned in the preceding chapters. However, there is a fourth assumption which is typical of multiple linear regression analysis: the linearity assumption. It states that the relationship between the dependent variable and the independent variables is linear. This assumption is automatically met when the independent variables are treated as nominal variables by means of effect or dummy coding, because the categories are compared to the mean score (effect coding) or to the last category (dummy coding). There are no assumptions on the kind of relationship between categories. Checking the linearity assumption becomes important if one or more of the independent variables have an interval or ratio level of measurement.

The same applies when interactions between independent variables are investigated, in which one of the interacting variables has an interval or ratio level of measurement. Table 7.15 gives data for an independent variable *X* and a dependent variable *Y*. It is evident that there is no linear relationship between the two variables. If *X* equals 2, the highest observations on *Y* are found. Let us assume that variable *X* represents three groups of speakers with a different language background and *Y* a variable which measures the ability to distinguish between specific types of vowels. One may have high expectations about the kind of speaker who performs well, but that does not imply that the relationship between group membership (the independent variable) and the performance on vowel discrimination (the dependent variable) is linear. First and foremost, the investigator is interested in the question whether meaningful differences between the groups occur or not. How can a variable with a low measurement level be processed in regression analysis?

One method of processing nominal or ordinal independent variables is the application of effect coding. In Table 7.15 effect coding is applied to the two factors of Table 7.12, 'network' and 'gender'. It is obvious how these two effect codings are obtained. The first variable (E1) marks the membership of a group labelled by the first factor ('network' low = 1, 'network' high = -1). The same procedure is used for the factor 'gender' (E2). Interaction is coded by the multiplication of E1 and E2: INT.

For example, we would like to assess the effect of the factor 'gender' We set up two sets of regression equations with effect coding: E1 represents the 'network' factor, E2 the 'gender' factor and INT the interaction. Table 7.15 contains the data matrix.

The effect codes can be entered as independent variables in a regression analysis. The outcomes are shown in Table 7.16.

Table 7.15. Relevant parts of SPSS data matrix representing data of Table 7.12

network	gender	vara	E1	E2	INT
1	1	1.73	1	1	1
1	1	2.13	1	1	1
1	1	2.16	1	1	1
1	1	2.38	1	1	1
1	1	1.45	1	1	1
1	1	2.63	1	1	1
1	1	1.75	1	1	1
1	1	1.05	1	1	1
1	1	2.25	1	1	1
1	1	2.42	1	1	1
1	1	2.78	1	1	1
1	2	2.65	1	-1	-1
1	2	2.73	1	-1	-1
1	2	2.45	1	-1	-1
1	2	2.85	1	-1	-1
1	2	3.05	1	-1	-1
1	2	2.73	1	-1	-1
2	1	2.33	-1	1	-1
2	1	2.48	-1	1	-1
2	1	2.75	-1	1	-1
2	1	2.70	-1	1	-1
2	1	2.35	-1	1	-1
2	1	2.33	-1	1	-1
2	1	2.50	-1	1	-1
2	2	2.83	-1	-1	1
2	2	3.78	-1	-1	1
2	2	3.13	-1	-1	1
2	2	3.43	-1	-1	1
2	2	3.15	-1	-1	1
2	2	3.03	-1	-1	1
2	2	2.75	-1	-1	1
2	2	3.03	-1	-1	1
2	2	2.30	-1	-1	1
2	2	2.55	-1	-1	1
2	2	2.12	-1	-1	1
2	2	2.78	-1	-1	1
2	2	2.18	-1	-1	1

Below in Table 7.16 we give the results of two regression analyses - representing Type I and Type III sums of squares - carried out on the data of Table 7.15.

Table 7.16. Two regression equations with different solutions for the Milroy (1980) data; E1 and E2 refer to the effect codes, and INT to the coding of the interaction

Set I	*SS* due to regression		Set II	*SS* due to regression	
E1, E2	3.774		E1,E2,INT	3.985	
E1	1.619		E1,INT	1.736	
Effect of E2 (Type I)	2.155	$R(\beta\|\mu,\alpha)$	Effect of E2 (Type III)	2.249	$R(\beta\|\mu,\alpha,\alpha\beta)$

In Set I the *SS* associated with E2 is obtained by first setting up a regression equation with the predictors E1 and E2 (corresponding to NETWORK and GENDER respectively) and the constant μ: $X = 2.514 - 0.135 \times E1 - 0.253 \times E2$, and calculating the *SS* associated with that equation (*SS* = 3.774). After that the *SS* associated with E1 is obtained by using E1 and μ as predictors ($X = 2.514 - 0.210 \times E1$: *SS* = 1.619. Subtracting the latter *SS* from the former we get the *SS* of the analysis of variance labelled Type I.

If we first set up a regression equation with E1, E2, INT (the interaction term) and μ as predictors, we obtain an *SS* of 3.985. Subsequently we subtract the *SS* associated with the regression equation containing E1, INT, and μ as predictors: 3.985 - 1.736 = 2.249. The result corresponds with the *SS* of GENDER in the type III analysis of variance (2.250; difference due to rounding errors).

Labels like $R(\beta|\mu,\alpha,\alpha\beta)$ become more obvious now. This label for instance means: the effect of factor B (here: GENDER), taking into account the effects of factors α (here NETWORK, represented by E2), $\alpha\beta$ (here the interaction NETWORK $\times$ GENDER, represented by INT) and μ). The isolation of the *SS* associated with E2 is here obtained by first calculating the *SS* due to regression with all three factors (E1, E2, INT and a constant) and subtracting from it the *SS* due to regression with E1 and INT as predictors: *SS* (Const, E1, E2, INT) – *SS* (Const, E1, INT): 3.985 – 1.736 = 2.249.

For illustrative purposes we carried out the regression analyses associated with some of the four different ways of obtaining Sums of Squares. The GLM model in SPSS (and SAS etc.) carries out this regression analysis when the appropriate options are chosen. In Table 7.17 the output of the SPSS procedure GLM: Univariate (option SS Type I) is printed; the *SS* values for GENDER are the same as obtained by the regression approach.

Table 7.17. SPSS output analysis of variance for the data from Milroy (1980), see Tables 7.14 and 7.20; *SS* Type I was used

Source	Type I Sum of Squares	df	Mean Square	F	Sig.
Corrected Model	3.985[a]	3	1.328	7.392	.001
Intercept	237.238	1	237.238	1320.132	.000
NETWORK	1.619	1	1.619	9.007	.005
GENDER	2.155	1	2.155	11.984	.001
NETWORK*GENDER	.211	1	.211	1,176	.286
Error	5.930	33	.180		
Total	247.154	37			
Corrected Total	9.916	36			

a. R Squared = .402 (Adjusted R Squared = .348)

7.7. Suggestions for reporting statistics

EXAMPLE 1

An analysis of variance (one-way, 'condition' factor fixed) was carried out on the log-transformed Hz values associated with the targets of the H*L pitch accents. Levene's test was significant ($F_{4.45} = 7.20$, $p < 0.05$); consequently the assumption of equality of variances was not met. We decided to apply a logarithmic transformation on the data. This transformation is ecologically valid, as the perception of intonation is based on log-transformed Hz-values (cf. Nolan 2003). The *F* ratio to test whether the conditions affected the targets of the pitch accents was significant: $F_{4,45} = 6.10$, $p < 0.05$; $\eta^2_{\text{partial}} = 0.045$.

7.8. Terms and concepts

- *Arcsine transformation*: a transformation used in analysis of variance especially when proportions constitute the dependent variable.
- *Effect coding*: a method to code nominal variables with –1 and +1. Codes for interactions are obtained by multiplication of the codes.

- *Homoscedasticity*: assumption of equal error variances in analysis of variance and multiple regression.
- *Unbalanced design*: a multi-factor design in analysis of variance with unequal cell sizes.

7.9. Exercises

1. Apply Levene's test to the data of Table 7.8 to test the assumption of homoscedasticity.
2. Compare the effects of the following transformations: $X = 2\arcsin\sqrt{p}$, and $X = \arcsin\sqrt{S/(N+1)} + \arcsin\sqrt{(S+1)/(N+1)}$ to the data given in Table 7.8, assuming that $N = 100$. Compare their effects by determining the ratios of the largest and smallest variances in both cases.
3. An experiment was carried out in order to assess the results of three literacy methods: Methods 1, 2 and 3. Five children were randomly assigned to one of the three methods. The following table gives the scores obtained after a final reading test; the scores range from 1 to 10.

M_1	M_2	M_3
3	9	4
2	10	10
4	10	7
6	10	3
5	10	1

   (a) Inspect the outcome of Levene's test.
   (b) Carry out an analysis of variance on the data.
   (c) Use Welch's test instead of an ordinary one-way analysis.
   (d) What might be the reason that the results achieved with method 2 differ so much from the ones achieved in Methods 1 and 3?
4. In which situation will the transform $\log(1+x)$ be more appropriate than $\log x$ when variable x is submitted to analysis of variance?
5. The following data set with unequal cell frequencies (see cell $AB_{2.2}$) is analyzed:

	B_1	B_2
A_1	1	2
	3	4
A_2	7	5
	6	

Which hypothesis is tested when $\bar{X}_{.1.} - \bar{X}_{.2.}$ is calculated?

6. Use a computer program to apply the regression approach (Method II) to the data of Milroy (1980) (see Table 7.12). Include both independent variables (gender and network) and their interaction. Calculate the sums of squares and test the significance of the effects.

Chapter 8
Repeated measures designs

8.1. Preview

Repeated measures designs are quite popular, because they restrict the required number of participants to a considerable extent as the same participant is submitted to a number of treatments; they constitute a so-called *within-subject factor*. A large number of subdisciplines in linguistics use these designs. In language and speech pathology the performance of patients is measured at several moments in time, for instance before therapy, and six and twelve months after therapy. In applied linguistics students are tested at regular intervals in order to assess whether they make progress in specific assignments, while in psycholinguistics or cognitive neurosciences the participants may be asked to carry out tasks in different conditions, with ERP measures or reaction times as dependent variables. In sociolinguistics, data may be obtained from the same participants in different stylistic conditions. In speech technology repeated measures occur too, for instance when listeners are asked to assess the quality of utterances generated by different systems for speech synthesis.

The analysis of data obtained in repeated measures designs is not straightforward. A number of approaches are possible, varying from the conventional univariate and multivariate analysis of variance to multilevel analyses, a technique which will be discussed in the next chapter. In this chapter we focus on analysis of variance. In Section 8.2 two types of factors are introduced: *within-subject factors* and *between-subject factors*; the latter distinguish groups of participants. In Sections 8.3 to 8.5 current types of repeated measures designs are discussed together with indices for important assumptions in repeated measures designs: *sphericity*, and the absence of interaction between participants and treatments. Psycholinguistic research frequently uses *hierarchical designs*: designs in which one factor (for instance 'words') is nested within a factor like 'type of words'. Difficulties arise especially if the factor 'words' is a random factor; solutions are discussed in Section 8.6. Section 8.7 discusses an alternative approach to repeated measures: *multivariate analysis* and the pros and cons of univariate and multivariate approaches. Although this book focusses on designs with one dependent variable, in Section 8.8 we give a short introduction to the analysis of data obtained on more than one dependent variable. In Sections 8.9 and 8.11 we present examples of variants of repeated measures designs with more than one within-subject factor or a mixture of a between-subject and a within-subject factor. In Section 8.10 we discuss post hoc comparisons in repeated measures designs. We devote a special section – Section 8.12 – to ways of dealing with missing data, a serious problem, especially in the analysis of data obtained in repeated measures designs.

8.2. Properties of within-subject designs

In repeated measures designs the same participant is submitted to a number of treatments; these treatments constitute the *within-subject factor.* If only one factor is involved (the randomized block design), such a design can be regarded as a two-factor design (treatments × subjects (participants)), with only one observation per cell.

Randomized block designs, including designs with *repeated measures* are characterized by the use of so-called blocks. The participants or participants are distinguished by these blocks in such a way that the units within a block are as similar as possible with respect to the dependent variable while the units in different blocks are less similar. Homogeneity within blocks can be achieved by using (1) litter mates or identical twins, (2) matched participants (e.g. matched by test scores) (3) repeated measures on a participant. The structure of a randomized block design is illustrated in Table 8.1.

Table 8.1. Layout of a one-way randomized block design; repeated measures occur if each participant constitutes a block

	Levels factor A			
	A_1	A_2	A_3	
Block 1	X_{11}	X_{12}	X_{13}	$\bar{X}_{1.}$
Block 2	X_{21}	X_{22}	X_{23}	$\bar{X}_{2.}$
Block 3	X_{31}	X_{32}	X_{33}	$\bar{X}_{3.}$
⋮	⋮	⋮	⋮	⋮
Block *n*	X_{n1}	X_{n2}	X_{n3}	$\bar{X}_{n.}$
	$\bar{X}_{.1}$	$\bar{X}_{.2}$	$\bar{X}_{.3}$	$\bar{X}_{..}$

The choice of a design has direct consequences for the elements or components of the equation model and for the way the variance is partitioned.

The model equation for a one-way randomized block design is the following:

$$\boxed{X_{ij} = \mu + \alpha_j + \pi_i + \varepsilon_{ij} \qquad (1)}$$

with α_j as the effect of treatment j of factor A, and π_i the effect associated with block or participant i. In comparison with a one-way randomized factorial design, the component π_i constitutes an additional source of variation.

When participants are assigned to all levels of a factor, or when blocks are used, the scores in the subpopulations formed by the levels of the factor in question are not independent anymore. This independence is assumed in the completely randomized factorial designs discussed in preceding chapters. In fact, a situation arises which is already known from the t tests for paired or correlated samples (cf. Chapter 3). We have to include a 'subject' or 'block' factor in our score model, in order to account for all the relevant variance components. The variation in a one-way randomized design with repeated measures can be subdivided as follows:

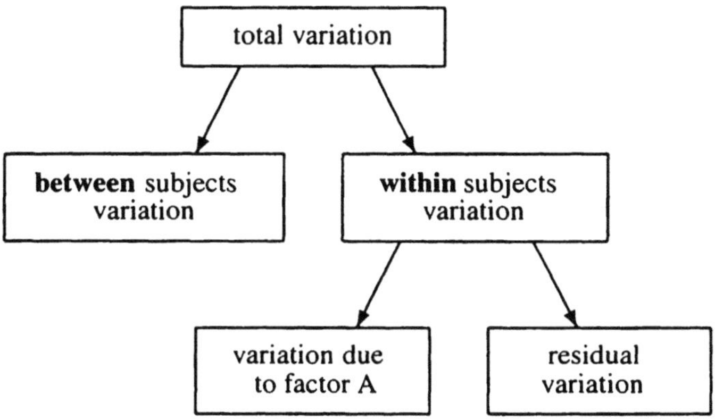

The subdivision or partition of the total variation as presented above shows the advantage of this design type. The between-subjects component can be separated from the within-subjects component. Neglecting the between-subjects component leads to a much larger error: The between-subjects variation is taken as part of the 'error', which is the case in a one-way completely randomized factorial design for participants within groups (see also Chapter 3 for paired samples in t tests). In a one-way randomized block design three variance components can be defined and traced, in a one-way completely randomized factorial design only two. However, it is important to note that in comparison to a two-way randomized factorial design, the interaction between the two factors is absent in the model equation of a randomized block design. As there is no within cell variation, the interaction $(\alpha\pi)_{ij}$, plus the error is assembled in the so-called 'residual variation'. These two sources of variation cannot be retrieved. In testing the effect α_j it is assumed that no interaction effect exists between participants (or blocks) and α_j. This assumption is discussed in Section 8.4.

Table 8.2 gives the expected values of the mean squares of a one-way randomized block design. The expected values given hold for the configurations occurring most frequently. In most cases the block effect is random when a randomized block design is used, whereas the other factor is fixed.

Table 8.2. Expected values of mean squares in a one-way randomized block design with A as fixed and the blocks (B) as random effect

Randomized block design		
MS	*df*	Expected Value
MS_A	$p-1$	$\sigma_\varepsilon^2 + \sigma_{\alpha\pi}^2 + n\sigma_\alpha^2$
MS_B	$n-1$	$\sigma_\varepsilon^2 + p\sigma_\pi^2$
MS_{res}	$(p-1)(n-1)$	$\sigma_\varepsilon^2 + \sigma_{\alpha\pi}^2$

Table 8.2 also shows how the effects can be tested. In a one-way randomized block design the effect α is tested against the residual error. It is worthwhile drawing attention to the possible advantage of a randomized block design (RBD) with repeated measures on participants over a completely randomized factorial design (CRF) using different groups of participants, who are randomly assigned to the treatments. To illustrate the differences, we compare a one-way design with three treatments and three groups of ten subjects to a design with three repeated measures on ten participants:

one-way factorial design	repeated measures
– A great number of participants, k times n_j, 30 in this example. – Differences between participants are included in the error variance.	– A restricted number of participants, 10 in this example. – The variance due to differences between participants is isolated in the 'block' effect; the resulting error (residual) is smaller than in the one-way factorial design. Consequently, the F test for the treatment effect is more powerful.
– The number of degrees of freedom for the error term is $N-k$; here $30-3=27$.	– The number of degrees of freedom for the error term is $(k-1)(n-1)$; here $2\times 9=18$, which is lower than in the one-way factorial design.

Thus, the advantages of an RBD are dominant, with a small disadvantage of the smaller number of *df* for the error term. Of course, it is not difficult to think of situations in which it is not possible or desirable to have participants subjected to different treatments, as effects of learning or fatigue may arise. Obviously these aspects are beyond the scope of this book.

The design of an experiment has consequences for the calculation of the mean squares, the degrees of freedom, the expected values of the mean squares and the F ratios. It is important, therefore, to be able to specify the nature of the design to be analyzed *precisely*. There is a wide variety of designs in which data can be collected. The way more complicated designs can be worked out and analyzed is explained in standard textbooks, such as Winer, Brown, and Michels (1991) and Kirk (1995). One frequently occurring design type should be mentioned here. It is a design which is a combination of a completely randomized factorial design and a repeated measures design; the result is a so-called *split-plot* design. Here the block units or repeated measures are restricted to a subset of the factors investigated, whereas the blocks (or participants) are distinguished in separate groups by at least one other factor. The latter factor is called the *between-subject factor*. This design type is illustrated in Table 8.3.

Table 8.3. Layout of a split-plot design; the blocks are distinguished in separate groups by between-subject factor A; factor B is measured in all blocks or participants

		B_1	B_2	B_3
	Block 1	X_{111}	X_{112}	X_{113}
A_1	Block 2	X_{211}	X_{212}	X_{213}
	⋮	⋮	⋮	⋮
	Block n	X_{n11}	X_{n12}	X_{n13}
	Block $n+1$	$X_{(n+1)21}$	$X_{(n+1)22}$	$X_{(n+1)23}$
A_2	Block $n+2$	$X_{(n+2)21}$	$X_{(n+2)22}$	$X_{(n+2)23}$
	⋮	⋮	⋮	⋮
	Block $2n$	$X_{(2n)21}$	$X_{(2n)22}$	$X_{(2n)23}$

As is illustrated in Table 8.3, the split-plot design is also marked by a hierarchical relationship between two factors. The block units are nested within factor A. This is a consequence of an aspect that has not been mentioned so far: As a matter of fact participants in a completely randomized factorial design are always nested under the factors involved. One may put the participant (subject) index in a model equation between brackets to indicate this nested relationship, (e.g. $\varepsilon_{i(jk)}$ instead of ε_{ijk} as is done in Kirk (1995)).

8.3. A univariate analysis of a repeated measures design

Assume that a speech and language pathologist wants to assess the progress made by five patients with a velopharyngeal dysfunction, i.e. the velum does not close sufficiently at the right time. These patients are measured with the nasometer, which gives the ratio of nasal sound pressure and the sum of nasal + oral sound pressures. The nasality of the patients is measured three times, at T_1, T_2 and T_3 (the values of nasality, called the 'nasalance scores', vary between 0 and 100).

Table 8.4. Nasalance scores of five participants at $T1$, $T2$ and $T3$

	T_1	T_2	T_3
S_1	50	40	40
S_2	40	41	40
S_3	42	35	36
S_4	47	42	44
S_5	52	46	47

Using the GLM, Repeated Measures procedure we obtain an output consisting of a large number of items. To begin with only the items reproduced in Table 8.5 are relevant.

In Table 8.5, we see four different significance values (①, ②, ③ and ④) associated with the same sums of squares and F ratios, quite a complicated result at first sight. In randomized block designs with k treatments and n blocks, the ratio of mean squares (the F ratio) is distributed as F with $df_1 = k - 1$, and $df_2 = (k-1)(n-1)$ if for all pairs of treatments the variances of differences between the levels of the within-subject factor (see Section 8.4) are equal. This assumption is the weaker version of the requirement that both variances of each of the measures and the covariances (or correlations) between the measures under all treatment levels should constitute homogeneous sets: the condition of *compound symmetry*, which is discussed in the next section.

Table 8.5. SPSS output of the GLM, Repeated Measures procedure for the data of Table 8.4; details will be discussed at the end of Section 8.4

Source		Type III Sum of Squares	df	Mean Square	F	Sig.
TIME ①	Sphericity Assumed	87.600	2	43.800	8.396	.011
③	Greenhouse-Geisser	87.600	1.119	78.272	8.936	.037
②	Huynh-Feldt	87.600	1.248	70.180	8.936	.031
④	Lower-bound	87.600	1.000	87.600	8.936	.044
Error (TIME)	Sphericity Assumed	41.733	8	5.217		
	Greenhouse-Geisser	41.733	4.477	9.322		
	Huynh-Feldt	41.733	4.993	8.359		
	Lower-bound	41.733	4.000	10.433		

Source		Partial Eta Squared	Noncent. Parameter	Observed[a] Power
TIME	Sphericity Assumed	.677	16.792	.860
	Greenhouse-Geisser	.677	9.397	.636
	Huynh-Feldt	⑤ .677	10.480	.681
	Lower-bound	.677	8.396	.591
Error (TIME)	Sphericity Assumed			
	Greenhouse-Geisser			
	Huynh-Feldt			
	Lower-bound			

a. Computed using alpha =.05

8.4. Assumptions in repeated measures designs

In order to explain the details presented in the SPSS output, we need to discuss the concepts of compound symmetry and Huynh-Feldt's adjustments of the degrees of freedom and we will do so on the basis of two data sets. For clarity's sake, we start with the stricter assumption. Assume we have the two data sets which are presented in Table 8.6, each consisting of a within-subject factor of three levels and four participants.

Table 8.6. Two data sets for a randomized block design with repeated measures

	data set A			data set B		
	T_1	T_2	T_3	T_1	T_2	T_3
participant 1	2	4	12	6	8	10
participant 2	3	3	5	3	3	5
participant 3	2	4	7	2	4	7
participant 4	2	4	11	2	4	5

Variances and covariances of the two data sets in Table 8.6 can be set out in a so-called covariance matrix C (see Appendix B):

$$C = \begin{pmatrix} s_1^2 & r_{12}s_1s_2 & r_{13}s_1s_3 & \cdots & r_{1k}s_1s_k \\ r_{21}s_2s_1 & s_2^2 & r_{23}s_2s_3 & \cdots & r_{2k}s_2s_k \\ r_{31}s_3s_1 & r_{32}s_3s_2 & s_3^2 & \cdots & r_{3k}s_3s_k \\ \cdots & \cdots & \cdots & \cdots & \cdots \\ r_{k1}s_ks_1 & r_{k3}s_ks_2 & r_{k3}s_ks_3 & \cdots & s_k^2 \end{pmatrix}$$

with s_j^2 = the variance of the observations in treatment j, and r_{ij} = the correlation between the observations on two treatments. A covariance matrix has compound symmetry if all of the variances on the main diagonal are equal and if all correlation coefficients are equal, too. The covariance matrices for both data sets of Table 8.6 are given in Table 8.7.

Table 8.7. Covariance matrices for the two data sets in Table 8.6

	data set A			data set B		
	T_1	T_2	T_3	T_1	T_2	T_3
T_1	.25	–.25	–1.25	3.58	3.75	3.75
T_2		.25	1.25		4.92	4.92
T_3			10.92			5.58

Clearly, the covariance matrix for data set B is much more homogeneous than that of data set A. The heterogeneity in the matrix can be caused either by different correlation coefficients or by heterogeneous variances (or both). Heterogeneous correlation coefficients may arise, for instance, when measures are taken at different time intervals; measures at adjacent moments in time tend to be more strongly correlated than measures obtained at moments further apart.

What is the consequence of not meeting the assumption mentioned above? Box (1954) showed that a violation of the assumption results in an increase of the real α level. In other words, hypotheses are rejected too often. The assumption of compound symmetry, however, is a sufficient but not a necessary condition. A less restrictive one is the *sphericity condition*, also called 'circularity assumption'. As already mentioned above, this condition requires the variances of all differences to be equal: $\sigma^2_{(x_i - x_j)}$ = constant. The condition can even be fulfilled if the matrix does not have compound symmetry (for an example see Winer et al.). This assumption is not tested on the original measures, but on derived ones obtained by transforming k repeated measures to $(k-1)$ new variables. The way in which these variables are obtained is comparable to the procedure we want to describe in this section: the multivariate approach. These new variables are obtained by setting up linear combinations of the differences between the original ones. The differences are assigned coefficients which ensure that the new variables are uncorrelated (orthogonal). A normalization of the coefficients is performed which ensures that the sum of squared coefficients is 1. The variables obtained in this way are said to form an 'orthonormal' set of variates. The sphericity condition requires these orthonormal variates to have equal variances. Obviously, a within-subjects factor with $df = 1$ will always satisfy the sphericity condition ($k-1=1$ in this particular case).

The index that expresses the extent to which the covariance matrix deviates from the ideal form is the index ε. There are two estimates of this index: 'Greenhouse-Geisser epsilon' and 'Huynh-Feldt epsilon'. The range of ε is:

$$1/(k-1) \leq \varepsilon \leq 1 \quad \text{with } k = \text{number of treatments} \qquad (2)$$

In our case the minimal value of ε is 0.5. If ε is close to its minimum value, we should reject the assumption of sphericity for the data involved. Next we can use ε to adjust the degrees of freedom of the F ratio as follows:

$$df_1 = \varepsilon \times (k-1) \qquad (3)$$

$$df_2 = \varepsilon \times (k-1)(n-1) \qquad (4)$$

By this reduction of df the F test becomes more conservative; we reduce the probability of committing a type I error. For data set A the following values of ε were obtained:

GREENHOUSE-GEISSER ε	=	0.52711
HUYNH-FELDT ε	=	0.56966

For data set B the following values were returned:

GREENHOUSE-GEISSER ε	=	0.78125
HUYNH-FELDT ε	=	1.00005

Both values for data set A approximate the minimum value of 0.5 quite closely.

By adjusting the degrees of freedom after Greenhouse-Geisser we get:

$$\begin{array}{rcl} df_1 = 0.5271 \times 2 & = & 1.054 = 1 \\ df_2 = 0.5271 \times 2 \times 3 & = & 3.163 = 3 \end{array}$$

The exact *df* values can be found in Table 8.8. At the 5% level the critical value of $F_{1,3}$ is 10.13. Thus the adjustment of the degrees of freedom for our F value of 11.91 results in an adjustment of the significance level, from 0.008 to just below 0.05. This can be observed in Table 8.8 as well.

Table 8.8. Part of the SPSS output of the GLM, Repeated Measures procedure for the two data sets of Table 8.6

data set A		Type III Sum of Squares	df	Mean Square	F	Sig.
TIME ①	Sphericity Assumed	92.667	2	46.333	11.914	.008
	Greenhouse-Geisser	92.667	1.054	87.901	11.914	.037
②	Huynh-Feldt	92.667	1.139	81.335	11.914	.032
	Lower-bound	92.667	1.000	92.667	11.914	.041
Error (TIME)	Sphericity Assumed	23.333	6	3.889		
	Greenhouse-Geisser	23.333	3.163	7.378		
	Huynh-Feldt	23.333	3.418	6.827		
	Lower-bound	23.333	3.000	7.778		

data set B		Type III Sum of Squares	df	Mean Square	F	Sig.
TIME ③	Sphericity Assumed	24.667	2	12.337	22.200	.002
	Greenhouse-Geisser	24.667	1.563	15.787	22.200	.006
④	Huynh-Feldt	24.667	2.000	12.333	22.200	.002
⑤	Lower-bound	24.667	1.000	24.667	22.200	.018
Error (TIME)	Sphericity Assumed	3.333	6	.556		
	Greenhouse-Geisser	3.333	4.888	.711		
	Huynh-Feldt	3.333	6.000	.556		
	Lower-bound	3.333	3.000	1.111		

The results of the analysis of variance of the two data sets in Table 8.6 are given in Table 8.8. We can see the following in Table 8.8:

① The output for data set A gives the ordinary degrees of freedom and significance levels.

② The Huynh-Feldt adjusted values are given as well. While for data set A the adjustment is quite considerable as a result of the small value of ε the probability value of the F ratio increases from 0.008 to 0.032 (i.e. the F ratio becomes less significant).

③ The ordinary significance level F ratio for the second data set is 0.002.

④ The 'ordinary' significance level of 0.002 is not affected by Huynh-Feldt, due to the fact that $\varepsilon = 1$.

⑤ Only if one does not use the estimated value of ε, but its minimal value or lower-bound to adjust the df, the significance level of F decreases too.

Returning to Table 8.5 we see the following:

① When sphericity is assumed incorrectly, $F_{2,8} = 8.396, p = 0.011$.

② Huynh-Feldt's adjustment gives 1.248 degrees of freedom, instead of the original 2, mentioned under 'sphericity assumed'. It implies that the value of ε for the nasalance scores of Table 8.4 was 0.624. This value is quite close to its minimum value of 0.500, thus we should reject the assumption of sphericity for the nasalance data, and report the Huynh-Feldt results. The effect of the 'Time' factor is significant at the 5% level for all adjustments of the degrees of freedom.

③ The Greenhouse-Geisser adjustment often yields results which are similar to those adjusted by Huynh-Feldt. This applies here, too.

④ Under the heading 'Lower-bound' we find the lowest possible value of the adjusted degrees of freedom. Consequently, the resulting probability value is higher than the others.

⑤ The effect size is quite considerable:
$\eta^2_{partial} = 0.667$ (= 87.600/(87.600 + 41.733)).

8.5. The interaction between participants and a within-subject factor

As only one observation per cell is available in repeated measures designs, the error component cannot be estimated by the $MS_{\text{within cells}}$. The interaction Treatments × Subjects is used instead; so the assumption is that the interaction behaves like 'error'. That is to say, it is expected to randomly vary from treatment to treatment. We illustrate this assumption with the two data sets given in Table 8.6.

The scores obtained for participants 1 and 4 in data set A exhibit different tendencies across the treatments from the scores obtained for participants 2 and 3, as can be clearly seen in Figure 8.1. In other words, there is an interaction between participants and treatments. Consequently, the associated *MS* cannot be regarded as an estimate of mere 'error'. In an ideal situation

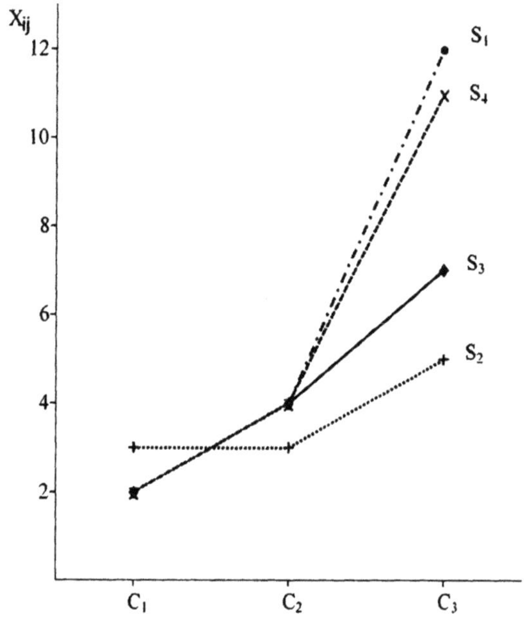

Figure 8.1. Participant × treatment plot of the data in data set A of Table 8.6

the Subject × Treatment plot has a form with curves that only differ in the intercept, in other words, the curves should be parallel.

When carrying out an F test for the treatment factor in a fixed-effects randomized block design, we have to assume that there is no interaction between blocks and treatments. If this assumption is not warranted, the value of MS_{res} – the error term for $MS_{\text{treatments}}$ – may be inflated, and consequently the corresponding F test is negatively biased. Tukey has developed a procedure to test the assumption of absence of interaction; this procedure is known as *Tukey's test of additivity*. If this test yields a significant F ratio, we have reason to believe that the 'Blocks' and 'Treatments' effects are not additive, but that they interact. Consequently, the expected value of MS_{res} is not σ_ε^2 but $\sigma_\varepsilon^2 + \sigma_{\alpha\pi}^2$. An increase of the expected value of the denominator of the F test obviously leads to a negative bias of that test. As a consequence, the underlying causes for a possible insignificant F ratio cannot be traced. We do not know, then, whether it is due to the absence of the effect in the population or to the negative bias in the test. For computational details we refer to Kirk (1995).

In Table 8.9 we present the output of the SPSS Scale, Reliability Analysis procedure for data set A in Table 8.6. The outcomes are generated by clicking on 'Tukey's test of additivity' under 'Statistics'.

Table 8.9. ANOVA table for repeated measures on one factor, applied to data set A in Table 8.6, SPSS Scale, Reliability Analysis procedure; source 'between people' refers to the participants, and 'between measures' to the 'Treatment' factor; NONADDITIVITY refers to Tukey's test

ANALYSIS OF VARIANCE					
SOURCE OF VARIATION	SUM OF SQUARES	DF	MEAN SQUARE	F	PROB
BETWEEN PEOPLE	10.9167	3	3.6389		
WITHIN PEOPLE	116.0000	8	14.5000		
BETWEEN MEASURES	92.6667	2	46.3333	11.914	.008
RESIDUAL	23.3333	6	3.8889		
NONADDITIVITY ①	22.6763	1	22.6763	172.565	.000
BALANCE	.6570	5	.1314		
TOTAL	126.9167	11	11.5379		
GRAND MEAN	4.9167				

TUKEY ESTIMATE OF POWER TO WHICH OBSERVATIONS ②
MUST BE RAISED TO ACHIEVE ADDITIVITY = −1.55

Table 8.9 gives the following information:

① If Tukey's test of (non)additivity is significant (in our example $F_{1,5}$ = 22.6763/0.1314 = 172.565), one may resort to a transformation of the data in order to achieve additivity.

① Here it is suggested to raise the observations to a power of −1.55. The Tukey test has only limited value in detecting interactional patterns; it is only sensitive to specific patterns.

8.6. Strange F ratios: testing hypotheses made difficult

Let us assume that a researcher has asked a number of participants to react to two types of words: type A and type B. Let us also assume that he/she has selected three words representative of each of the two word types. The dependent variable is the reaction time of the participants in milliseconds. There are four participants. The fictitious data set is presented in Table 8.10.

We want to test whether the participants react faster to type A words than

Table 8.10. Fictitious data set for a hierarchical repeated measures design (S = subject factor) with two factors: 'word type' (= T, two values), and 'word' (= W, six values, nested within 'word type')

	T_A			T_B		
	W_1	W_2	W_3	W_4	W_5	W_6
S_1	600	630	650	630	640	660
S_2	590	610	600	600	620	630
S_3	620	600	580	640	610	600
S_4	580	620	620	600	630	630

to type B ones. The participants, factor S, constitute a random effect, i.e. they are selected randomly. The factor 'word type' T is regarded as a fixed factor. The selected words for each of the two word types (= W(T)) could be a random or a fixed factor. The status of this last factor has far-reaching consequences in testing the effects. The expected values of the mean squares are given in Table 8.11. By inspecting this table one can find out whether a suitable error term is available to test the effect of word type (= T).

Table 8.11. Expected values of mean squares for the design of Table 8.10; (Q-q)/Q is 0 if the factor 'words' (= W(T)) is fixed, and 1 if this factor is random; S is the factor 'participants' (subjects), T the 'word type' factor

Source	*df*	expected value mean square
s	$r-1$	$\sigma_\varepsilon^2 + \frac{Q-q}{Q}\sigma_{SW(T)}^2 + pq\sigma_S^2$
t	$p-1$	$\sigma_\varepsilon^2 + \frac{Q-q}{Q}\sigma_{SW(T)}^2 + q\sigma_{ST}^2 + \frac{Q-q}{Q}r\sigma_{W(T)}^2 + qr\sigma_T^2$
w(t)	$p(q-1)$	$\sigma_\varepsilon^2 + \sigma_{SW(T)}^2 + r\sigma_{W(T)}^2$
st	$(p-1)(r-1)$	$\sigma_\varepsilon^2 + \frac{Q-q}{Q}\sigma_{SW(T)}^2 + q\sigma_{ST}^2$
sw(t)	$p(q-1)(r-1)$	$\sigma_\varepsilon^2 + \sigma_{SW(T)}^2$

If the selected words constitute a sample of a population of words, the factor W(T) is a random factor, and fraction $(Q-q)/Q$ in Table 8.11 is 1. The consequence, however, is that there is no appropriate error term to test the mean square of T, as you can verify for yourself. There is no other mean square which is equal to the mean square of T except for the component $qr\sigma_T^2$.

In SPSS, the GLM, Repeated Measures procedure nested within-subject factors cannot be specified. One has to resort to the MANOVA procedure (we refer to Page, Braver, and MacKinnon (2002) for further information on syntax). In our situation there is no appropriate error term for the 'type' factor if the nested factor 'words' is random. For a nested design with the within-subjects factors 'type' and 'word' (nested under 'type') the following syntax lines are needed, where w1 to w6 are the variables representing the six words:

```
MANOVA w1 w2 w3 w4 w5 w6
  /WSFACTORS= type(2) word(3)
  /WSDESIGN= type word WITHIN type
  /PRINT SIGNIF (AVERF HF).
```

The output file of the MANOVA procedure contains all relevant information, but one needs some expertise to retrieve it. The results are shown in Table 8.12.

Table 8.12. (Part of) ANOVA table produced by SPSS MANOVA (syntax mode) for the data set of Table 8.10; W(T) is assumed to be a fixed factor

```
Tests involving 'TYPE' Within-Subject Effect.

Tests of Significance for T2 using UNIQUE sums of squares
Source of Variation            SS    DF        MS         F      Sig of F

WITHIN+RESIDUAL  ①        12.50     3      4.17
TYPE             ②      1504.17     1   1504.17    361.00          .000
-------------------------------------------------------------------------
Tests involving 'WORD WITHIN TYPE' Within-Subject Effect.

Mauchly sphericity test, W =          .00000
Chi-square approx. =                  .        with 9 D. F.
Significance =                        .

Greenhouse-Geisser Epsilon =          .30196
Huynh-Feldt Epsilon =                 .39496
Lower-bound Epsilon =                 .25000
-------------------------------------------------------------------------
EFFECT .. WORD WITHIN TYPE
Tests involving 'WORD WITHIN TYPE' Within-Subject Effect.

AVERAGED Tests of Significance for w using UNIQUE sums of squares
Source of Variation            SS    DF        MS         F      Sig of F

WITHIN+RESIDUAL  ③      4700.00    12    391.67
WORD WITHIN TYPE ④      1033.33     4    258.33       .66          .632
```

We can only obtain an ANOVA table such as Table 8.12 if we can assume that the nested factor 'words' is a fixed factor.

Table 8.12 shows the following results:

① WITHIN+RESIDUAL is the error term obtained by subtracting the *SS*s associated with the factors 'subjects' (participants), 'type' and 'words within type' from SS_{total}. It is equivalent to MS_{ST} given in Table 8.11.

② When the 'words' factor is fixed – as assumed in SPSS – the error term mentioned under (1) is used and we obtain a significant effect for the 'type'factor: $F_{1,3} = 361.00$, $p = 0.000$.

③ WITHIN+RESIDUAL is the error term for the factor 'words within type'. It is equivalent to $MS_{SW(T)}$ as given in Table 8.11.

④ The 'words' factor is not significant: $F_{4.12} < 1$.

If 'words' has to be regarded as a random factor we obtain different results, depending on the approach which is chosen to deal with the fact that no appropriate error term is available for a conventional *F* ratio.

Some statisticians recommend the use of the so-called quasi *F* ratio: F' ('*F* prime') for situations in which no appropriate error term can be found (in our example with 'words' as random factor). The distribution of F' approximates the 'normal' *F* ratio and is obtained by combining mean squares in the numerator as well as in the denominator. The objective is, of course, to obtain an expected value of the numerator that exceeds that of the denominator only by the variance of the factor tested. In our case the proper F' is:

$$F' = \frac{MS_T + MS_{SW(T)}}{MS_{ST} + MS_{W(T)}} \qquad (5)$$

which corresponds to:

$$F' = \frac{\sigma_\varepsilon^2 + \sigma_{SW(T)}^2 + q\sigma_{ST}^2 + r\sigma_{W(T)}^2 + qr\sigma_T^2 + \sigma_\varepsilon^2 + \sigma_{SW(T)}^2}{\sigma_\varepsilon^2 + \sigma_{SW(T)}^2 + q\sigma_{ST}^2 + \sigma_\varepsilon^2 + \sigma_{SW(T)}^2 + r\sigma_{W(T)}^2} \qquad (6)$$

Numerator and denominator of this F' ratio are identical except for the variance component belonging to the effect T. $MS_{SW(T)}$ is WITHIN+RESIDUAL, marked by ③ in Table 8.12 and MS_{ST} corresponds with ① in this table. The value of F' in our example is:

$$F' = \frac{1504.20 + 391.67}{4.17 + 258.33} = 7.22$$

The calculation of the degrees of freedom of an F' is somewhat complicated, see formulas (7) and (8). MS_1 and MS_2 in these formulas refer to the two mean squares in the numerator of F', and MS_3 and MS_4 to those in the denominator; df_1, df_2, df_3 and df_4 refer to the associated degrees of freedom. The number of degrees of freedom for the numerator is the nearest integer value of:

$$df_{\text{num.}} = \frac{(MS_1 + MS_2)^2}{MS_1^2/df_1 + MS_2^2/df_2} \qquad (7)$$

The number of degrees of freedom for the denominator is the nearest integer value of:

$$df_{\text{den.}} = \frac{(MS_3 + MS_4)^2}{MS_3^2/df_3 + MS_4^2/df_4} \qquad (8)$$

These formulas give the following outcomes for the degrees of freedom in our example:

$$df_{\text{num.}} = \frac{3594323.1}{(1504.2^2/1) + (391.67^2/12)} = 1.57 \approx 2$$

$$df_{\text{den.}} = \frac{68904.518}{(4.1667^2/3) + (258.33^2/4)} = 4.13 \approx 4$$

$F'_{2.4} = 7.22$ turns out to be significant at the 5% level: 'word type' affects reaction time.

The merits of quasi F ratios have been the subject of some debate. In order to avoid Type I errors, and because the calculation of quasi F is only possible with complete data sets, Clark (1973) and Coleman (1979) recommended the use of the minimum value of F', min F', which is defined as:

$$\min F' = \frac{MS_T}{MS_{ST} + MS_{W(T)}} \qquad (9)$$

where the df of the numerator is the df of MS_T, and where the number of degrees of freedom of the denominator is calculated in the same way as for the plain F' ratio. The number of df is the nearest integer of:

$$df = \frac{(F1 + F2)^2}{F1^2/n_2 + F2^2/n_1} \qquad (10)$$

In our example min F' is 8.730, with $df_1 = 1$ and $df_2 = 4$; this outcome is *not* significant at the 5% level.

$F1$ and $F2$ in repeated measures designs

Alternative and very frequently used statistics – at least in (psycho)linguistics – are the so-called $F1$ and $F2$ tests. $F1$ refers to the test of the treatment factor ($= T$) by assuming that the participants constitute a random factor and the word material a fixed factor, whereas in the case of $F2$ treatment is tested by regarding participants as a fixed and words as a random factor. Very often both tests are combined in one decision rule: Both $F1$ and $F2$ have to be significant. The popularity of these tests in psycholinguistics is due to the fact that they can also be calculated on data sets with missing data. The trick is to pool over items ('words') to calculate $F1$. In our example the design is reduced to a simple randomized block design; the corresponding F is called 'over items', and to pool over participants in order to obtain $F2$ ('over participants' = 'over subjects'). In the latter case the design is reduced to a one-way design, with the 'treatment' factor, and the scores per items pooled over participants. See Tables 8.13a and 8.13b.

Table 8.13a. $F1$ design (S = 'subject' factor) with 'word type' (= T, two values) as within-subject factor

	T_1	T_2
S_1	627	643
S_2	600	617
S_3	600	617
S_4	607	620

Table 8.13b. $F2$ design: one-way design with factor 'word type' (= T, two values); the observations are obtained by pooling over participants/subjects

T_1	T_2
597	617
615	625
612	630

$F1 = 276.91$ ($df1 = 1$, $df2 = 3$), $p < 0.01$ and $F2 = 5.65$ ($df1 = 1, df2 = 4$), $p = 0.076$. It is not surprising that $F1$ is significant; in fact it is based on the wrong assumption that the items ('words') constitute a fixed factor. The variation between the two levels of factor T is to a large extent due to the random variation of the items.

Min F' can be calculated on the basis of $F1$ and $F2$:

$$\min F' = \frac{F1 \times F2}{F1 + F2} \tag{11}$$

In our example $\min F'_{1,4} = (276.91 \times 5.65)/(276.91 + 5.65) = 5.54, p > 0.05$. Both F' and min F' are known to be conservative tests (cf. Forster and Dickinson 1976), which means that in a number of cases researchers accept H_0 on the basis of these F ratios, although in fact the H_0 should be rejected. The use of $F1$ and $F2$ is questionable; the type I error associated with $F1$ is known to be large. $F2$ is questionable for two reasons:

- $F2$ is known to be sensitive to the interaction *ST* ('subjects (participants) by type'): If this interaction is present, the probability of committing a type I error for $F2$ is high. This interaction is a crucial one, both for methodological and statistical reasons. If there is an interaction between 'participants' and 'type' we must assume that Participants react differently to items of specific Types. Testing this interaction is only possible with complete data sets. So, if there are missing data, one should impute the missings (cf. Chapter 9).
- The crucial problem of $F2$ is that statistical analyses which ignore within subject correlations – as is the case with $F2$ – tend to give incorrect inferences (cf. Gornbein, Lazaro, and Little 1992).

In order to avoid unacceptably high Type II error rates, the *multistage decision rule* proposed by Forster & Dickinson (1976: 141) can be recommended.

Forster and Dickinson multistage decision rule

1. Test min F'. If it is significant reject H_0. If it is not significant, proceed to step 2.
2. Test effect of ST (interaction 'subjects by type') and W(T) ('words within type'). These effects are tested by dividing their mean squares by $MS_{SW(T)}$. If either of them produces a non-significant F, proceed to step 3. Otherwise accept H_0.
3. Test the so-called $F1$ and $F2$. $F1$ equals MS_T/MS_{ST} and $F2$ equals $MS_T/MS_{W(T)}$. If both F ratios produce a significant result, reject H_0; otherwise accept H_0.

As a variant of the procedure suggested by Forster & Dickinson, we propose the following procedure, in which *imputation* plays a role. Imputation is a term which refers to filling in missing data according to specific criteria. A number of procedures is available (see Chapter 9). If 'words' of our example is a genuine random factor, and for instance observation $X_{32(5)}$ is missing one may use data of the same subject (in our example subject 3), and the same level of the within-subject factor 'type of words' (in our example level 2). One could randomly draw an observation – with replacement – from cell X_{32} and replace the missing observation with it. This procedure is called a 'hot deck' approach, see Chapter 9.

our multistage decision rule

1. Are there missing data? Use an imputation procedure when appropriate (see Chapter 10).
2. Test the effect of ST (subjects by type interaction) and W(T). These effects are tested by dividing their mean squares by $MS_{SW(T)}$. If the effect ST is not significant, use $F2$.
3. If ST is significant, assess whether this interaction is disordinal; if this is the case, stop the analysis in terms of main effects.
4. If ST is an ordinal interaction and W(T) is not significant, use $F1$.
5. If both ST and W(T) are significant, use min F'. If it is significant reject H_0.

The quasi F ratio in relation to the sphericity condition has not yet been taken into consideration. Santa, Miller, and Shaw (1979) and Maxwell and Bray (1986) showed that this test statistic is robust to both violations of distribution and homogeneity of variance assumptions and violations of the sphericity condition. This implies that we should not refrain from using F' when a 'normal' error term cannot be found. One may even question the need to use alternatives like $F1$, $F2$ and min F'.

- Explain in your own words what is meant by the distinction between 'within' and 'between-subject' factors.
- Which problems arise when the within-subject factor A (nested in within-subject factor B) is a random factor?
- Is it possible to calculate a quasi F ratio when there are missing data in a within-subject design?
- Is a Huynh-Feldt adjustment of the degrees of freedom for a within-subject factor relevant when that factor has two levels? Why?
- What are the similarities of a split-plot design and a design which has a covariate?
- What does a significant 'test of additivity' mean in the analysis of a repeated measures design?
- What is assumed in an $F1$ and an $F2$ analysis? (viz. the distinction between 'random' and 'fixed' factors)

8.7. Multivariate analysis in repeated measures designs

A frequently used alternative to the univariate repeated measures analysis is the multivariate approach. The multivariate analysis for the one-sample case resembles the t test for paired observations. Remember that this t test has the following form for two treatments:

$$t = \frac{\sum D/N}{s_D/\sqrt{N}} = \frac{\bar{D}}{s_D/\sqrt{N}} \qquad (12)$$

In this formula $\sum D$ is the sum of the differences between the paired scores and s_D^2 the standard deviation of the differences. Squaring t the formula is as follows $((s_D^2)^{-1} = 1/s_D^2)$:

$$t^2 = N \times \bar{D} \times (s_D^2)^{-1} \times \bar{D} \qquad (13)$$

The multivariate counterpart for k treatments with $k-1$ difference vectors is 'Hotelling's T^2':

$$T^2 = N \times d' \times (s_d^2)^{-1} \times d \qquad (14)$$

The associated degrees of freedom are: $df_1 = k-1, df_2 = N-k+1$. This equation contains so-called vectors (d and d') and a matrix (s_d^2). It is clear that the equation for Hotelling's T^2 bears a strong resemblance to the simpler t^2. To understand the steps in the calculation of T^2 one should be familiar with vectors and matrices (cf. Appendix B). The vector d contains the means of the $k-1$ differences between the treatments. The F ratio corresponding to Hotelling's T^2 can be obtained as follows:

$$F = \frac{N-k+1}{(N-1)(k-1)} \times T^2 \qquad (15)$$

For data set A in Table 8.6 we obtain the difference vectors given in Table 8.14. The covariance between the two difference vectors is: $(\sum d_1 d_2 - (\sum d_1 \sum d_2/N))/(N-1) = (36-(-6 \times -20)/4)/3 = 2.00$. A third difference vector is possible: $d_3 = T_1 - -T_3$; this vector is not included because its value can be predicted by using the two other vectors: $d_3 = d_1 + d_2$. Given three measures, two difference vectors contain the relevant information.

Table 8.14. Difference vectors for data set A in Table 8.6; $d_1 = T_1 - T_2$, $d_2 = T_2 - T_3$

	d_1	d_2	$d_1 d_2$
	–2	–8	16
	0	–2	0
	–2	–3	6
	–2	–7	14
Σd	–6	–20	36
$\bar{d}$	–1.50	–5.00	
s_d^2	1.00	8.66	

With formula (11), we get the following outcome for Hotelling's T^2:

$$\begin{aligned} T^2 &= N \times (\bar{d}_1\ \bar{d}_2) \begin{pmatrix} s_1^2 & r_{12}s_1s_2 \\ r_{12}s_1s_2 & s_2^2 \end{pmatrix}^{-1} \begin{pmatrix} \bar{d}_1 \\ \bar{d}_2 \end{pmatrix} \\ &= 4 \times (-1.50\ -5.00) \begin{pmatrix} 1.00 & 2.00 \\ 2.00 & 8.66 \end{pmatrix}^{-1} \begin{pmatrix} -1.50 \\ -5.00 \end{pmatrix} \\ &= 4 \times (-1.50\ -5.00) \begin{pmatrix} 1.858 & -0.429 \\ -0.429 & 0.215 \end{pmatrix} \begin{pmatrix} -1.50 \\ -5.00 \end{pmatrix} \\ &= 12.4335 \end{aligned} \tag{16}$$

The associated $F = (4-3+1)/(3 \times 2) \times T^2 = 0.333 \times 12.4335 = 4.1404$; $df_1 = 2$ and $df_2 = 2$. F is not significant at the 5% level. Consequently, we reject the hypothesis that the conditions affect the dependent variable. In Table 8.15 we reproduce the SPSS output of the multivariate analysis in question, which is part of the SPSS output of the GLM, Repeated Measures procedure.

Table 8.15. SPSS GLM, Repeated Measures

Effect		Value	F	Hypothesis df	Error df	Sig.
FACTOR 1	Pillai's Trace	.806	4.143	2.000	2.000	.194
	Wilks'Lambda	.194	4.143	2.000	2.000	.194
	Hotelling's Trace	4.143	4.143	2.000	2.000	.194
	Roy's Largest Root	4.143	4.143	2.000	2.000	.194

Table 8.15 shows that several multivariate tests are available, e.g. Pillai's Trace, Wilks' Lambda, Hotelling's Trace, and Roy's Largest Root; we only discussed Hotelling's T^2. It may have been noticed that the Hotelling's statistic, as presented by SPSS MANOVA, differs from the one calculated by hand.

The corresponding F test, however, agrees with our F test apart from rounding errors: 4.143. The reason is that SPSS gives 'Lawley-Hotelling's trace criterion' whereas we calculated Hotelling's T^2. Both statistics can be related to the F distribution. The F value found is not significant. Table 8.8 gives the results of the univariate approach, which delivers a significant result. We must conclude, therefore, that the univariate and multivariate approaches led to contradictory results. The univariate F was significant at the 5% level, whereas the multivariate F was not significant at all ($p = 0.194$). Therefore, we would recommend the replication of the experiment with a larger number of observations.

Univariate vs. multivariate analysis

Summarizing the preceding section, we can say that there are two approaches to the analysis of repeated measures designs:

1. The univariate approach:
   - Calculate the F ratio for the within-subjects factor.
   - If this F ratio is not significant, stop the analysis.
   - If the F test is significant, adjust the corresponding degrees of freedom by multiplying them by ε, and carry out the F test(s) with the adjusted df.
2. The multivariate approach:
   - The multivariate approach is only possible if the number of participants (blocks) exceeds the number of levels of the within-subjects factor.
   - Use the F approximations of the multivariate tests of the within-subjects factor: Hotelling's T^2, Wilks' Lambda or the Pillai-Bartlett statistic (PILLAIS).

It is quite difficult to evaluate the pros and cons of the two approaches. Rogan, Keselman and Mendoza (1979) found that both approaches are comparable with respect to Type I error control and power (although there are also reports which point to a lower power of multivariate tests). In this section we have focussed on the conditions for the within-subjects factors in repeated measures designs. If the design also contains between-subjects factors, the assumption of homogeneity of variance for these factors has to be tested by Box's M statistic.

8.8. Genuine multivariate analysis

Another technique directly related to analysis of variance is multivariate analysis of variance, MANOVA. MANOVA is a technique which deals with more than one dependent variable. Standard ANOVA is univariate, with only one dependent variable. One can think of many situations in which more than one dependent variable is involved. In many sociolinguistic research projects more than one linguistic variable is studied and one can imagine that several dependent variables can be analyzed simulataneously in a multivariate extension of the univariate form of analysis of variance. An example is the duration, pitch and intensity (three dependent variables) of vowels as a function of vowel type and language community (two independent variables or factors). The price to be paid is the complexity of the outcomes. If there is a significant effect, which of the different dependent variables is affected? In many situations, separate univariate analyses of variance will give clear results. The multivariate approach is attractive when there is an evident relationship between the dependent variables.

A good example of an intrinsic relationship are the formant values (spectral peaks) of vowels. Formants are the strongest harmonics of the fundamental frequency of a vowel. They are the result of the resonances of the vocal tract, which vary as a function of the shape of the cavities of the vocal tract for the sound to be realized (for instance [i] vs. [a]; for [i] the tongue is in a higher and more frontal position than for [a]). The values of formant 1, F1, and formant 2, F2, are core characteristics of vowels; they are measured in Hz. In Table 8.16, a fictitious data set is given with the formant values of [i] of 12 speakers. The speakers belong to two social classes and the question is whether the two classes pronounce the [i] differently.

If separate univariate analyses of variance are applied to the F1 and F2 data of Table 8.16, the results are not significant. For the first formant the result is $F_{1,10} = 3.083, p = 0.112$, for the second $F_{1,10} = 4{,}043, p = 0.072$. To get a visual impression of the differences between the lower and higher social classes, it is helpful to plot the F1 and F2 values in a two-dimensional space, preferably one that matches the vowel space representation used in linguistic research. Such a plot shows a sharper distinction than one observes for the two formant values separately. This is confirmed by the outcomes of a MANOVA analysis of the data of Table 8.16. These outcomes can be found in Table 8.17.

Table 8.16. Data set of twelve [i]-like vowels produced by twelve speakers from two social classes; the first two formant values of the vowels were measured, F1 and F2

social class	speaker	F_1	F_2
lower	1	300	2420
	2	360	2480
	3	310	2490
	4	315	2500
	5	320	2380
	6	420	2510
higher	7	330	2340
	8	360	2350
	9	400	2440
	10	410	2480
	11	370	2380
	12	385	2420

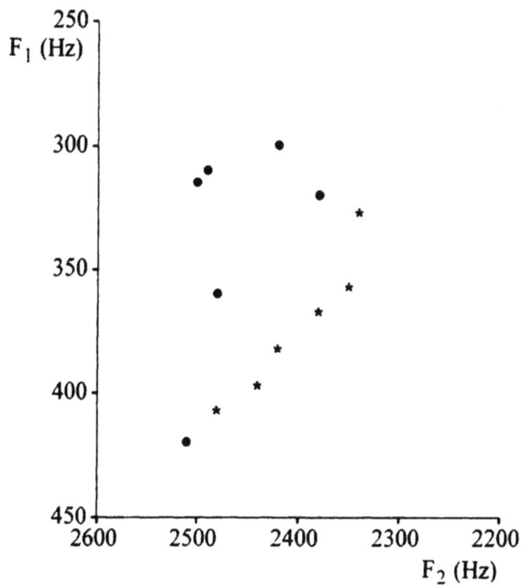

Figure 8.2. Plots of the data set of twelve [i]-like vowels produced by twelve speakers, for F1 (*y* axis) and F2 (*x* axis); • = lower class ⋆ = higher class

Table 8.17. SPSS output data Table 8.16, multivariate analysis of variance

Effect		Value	F	Hypothesis df	Error df	Sig.
Intercept	Pillai's Trace	1.000	1504.771	2.000	9.000	.000
	Wilks' Lambda	.000	1504.771	2.000	9.000	.000
	Hotelling's Trace	3422.616	1504.771	2.000	9.000	.000
	Roy's Largest Root	3422.616	1504.771	2.000	9.000	.000
Class	Pillai's Trace	.673	9.252	2.000	9.000	.007
	Wilks' Lambda	.327	9.252	2.000	9.000	.007
	Hotelling's Trace	2.056	9.252	2.000	9.000	.007
	Roy's Largest Root	2.056	9.252	2.000	9.000	.007

Several multivariate statistics are available with associated F values, as can be seen in Table 8.17. Wilks' Lambda (Λ) is frequently used and is defined as the ratio of two (determinants of) matrices: Matrix B (with sums of squares of the between-group scores), and Matrix E (with the error-variance); see Appendix B for these matrices and the concept of 'determinant':

$$\Lambda = \frac{|E|}{|B+E|} \tag{17}$$

In contrast to the F ratio smaller values of Λ signalize stronger effects of the factors. In our example $\Lambda = 0.327$, the associated $F_{2,9}$ is 9.252, $p = 0.007$. The conclusion is that the independent variable 'Class' makes a difference. There is a significant difference between the two social classes in the pronunciation of [i].

The following steps lead to a multivariate analysis of variance in SPSS:

window Data View	*click on*	Analyze
	go to	General Linear Model
	go to +	
	click on	Multivariate
window Multivariate		
the dependent variables:	*click on*	the variables 'f1' and 'f2'
	insert	under dependent variables
the independent variables:	*click on*	the variable 'class'
	insert	under fixed factors
	click on	OK

Our multivariate example delivers the following syntax by clicking Paste:

```
GLM
  f1 f2 BY class
  /METHOD = SSTYPE(3)
  /INTERCEPT = INCLUDE
  /CRITERIA = ALPHA(.05)
  /DESIGN = vowel .
```

In many settings researchers not only want to see whether a factor significantly affects the position of the scores in the n-dimensional space (this is the space spanned by the n dependent variables). They also have specific hypotheses with regard to the effects of the factor on each of the dependent variables. The latter is the case if the dependent variables are 'conceptually independent' (Biskin 1980: 70). This often leads the investigator to carry out separate univariate analyses of variance on each of the n dependent variables with the same factors. There are a number of noteworthy aspects in this context:

- If a number of tests is carried out on the same group of participants we risk an unacceptable high Type I error probability (see Chapter 4, where we introduced this problem in the context of multiple t tests). One way to avoid this high Type I error is to apply the so-called Bonferroni adjustment. If there are k tests, the actual α level for test k (α_k) is: $\alpha_k \leq 1-(1-\alpha)^k < k\alpha$, in which α is the significance level to be achieved, very often 0.05. In practice one often applies the rule: Divide α by k. In the situation where there are two dependent variables this means: 0.05/2 = 0.025. In general the Bonferroni adjustment procedure is quite conservative.
- A significant effect obtained in MANOVA does not imply that any significant effect will be obtained in a univariate analysis of variance (Tatsuoaka 1971: 23).

8.9. Two within-subject factors

In most languages a stressed vowel, like in the English word ''permit' is longer than the unstressed counterpart, like in 'per'mit'. Assume a phonetician wants to know whether the 'Stress' factor affects the durations of vowels to the same extent. This boils down to the question whether there is an interaction between the factors 'Stress' and 'Vowel'. The phonetician asked ten speakers to realize three vowels – [i], [a] and [I] – in stressed and unstressed conditions. The data are given in Table 8.18.

The design is a repeated measures design with two crossed within-subject factors: 'stress' (2 levels) and 'vowel' (3 levels). SPSS assumes that both within-subject factors are fixed, and that the 'speaker' factor is random. If there is an interaction between 'vowel' and 'stress', we can assume that stress does not affect every vowel to the same extent. Below the data set we show the syntax needed for the analysis of this design.

Table 8.18. Vowel durations in ms as a function of stress (two levels) and vowel type (three levels) realized by ten speakers

	– stress			+ stress		
vowel	[i]	[a]	[I]	[i]	[a]	[I]
S_1	70	65	55	90	100	75
S_2	60	65	65	85	105	80
S_3	50	55	55	70	85	75
S_4	70	65	65	85	105	90
S_5	75	70	70	90	110	80
S_6	80	85	85	100	115	100
S_7	65	60	70	85	115	95
S_8	75	80	70	95	120	80
S_9	70	65	65	80	105	85
S_{10}	60	70	60	80	125	80

The syntax for this repeated measures design with two crossed within-subject factors is:

```
GLM v1 TO v6
  /WSFACTOR = stress(2) vowel(3)
  /PRINT = DESCRIPTIVE ETASQ OPOWER
  /DESIGN.
```

Table 8.19 shows the following:

① The assumption of sphericity was not violated as the Huynh-Feldt adjusted degrees of the factor 'Vowel' and the interaction 'stress × vowel' do not differ from the nominal degrees of freedom ('Sphericity Assumed').

② The results show a strong effect of the interaction 'Vowel × Stress': $F_{2.18} = 45.833$, $p < 0.001$, $\eta^2_{partial} = 0.836$. This means that the effect of stress is different for the three vowels involved.

Table 8.19. Part of the SPSS output of the GLM, Repeated Measures procedure for the data of Table 8.18

Source		Type III SS	df	Mean Square	F	Sig.
STRESS	Sphericity Assumed	9881.667	1	9881.667	407.336	.000
(1)	Huynh-Feldt	9881.667	1.000	9881.667	407.336	.000
Error	Sphericity Assumed	218.333	9	24.259		
(STRESS)	Huynh-Feldt	218.333	9.000	24.259		
VOWEL	Sphericity Assumed	2072.500	2	1036.250	19.931	.000
(1)	Huynh-Feldt	2072.500	2.000	1036.250	19.931	.000
Error	Sphericity Assumed	935.833	18	51.991		
(VOWEL)	Huynh-Feldt	935.833	18.000	51.991		
STRESS *	Sphericity Assumed	1650.833	2	825.417	45.833	.000
VOWEL (2)	Huynh-Feldt	1650.833	2.000	825.417	45.833	.000
Error	Sphericity Assumed	324.167	18	18.009		
(STRESS *						
VOWEL)	Huynh-Feldt	324.167	18.000	18.009		

Source		Partial Eta Squared	Noncent. Parameter	Observed[a] Power
STRESS	Sphericity Assumed	.978	407.366	1.000
	Huynh-Feldt	.978	407.366	1.000
Error (STRESS)	Sphericity Assumed			
VOWEL	Sphericity Assumed	.689	39.863	1.000
	Huynh-Feldt	.689	39.863	1.000
Error	Sphericity Assumed			
(VOWEL)	Huynh-Feldt			
STRESS *	Sphericity Assumed	.836	91.666	1.000
VOWEL	Huynh-Feldt	.836	91.666	1.000
Error	Sphericity Assumed			
(STRESS *	Huynh-Feldt			
VOWEL)	Huynh-Feldt			

a. Computed using alpha =.05

8.10. Post hoc comparisons

If the interaction 'stress × vowel' had not been significant, it would have been a valid question which of the three vowels constitute homogeneous sets. The SPSS GLM, Repeated Measures procedure provides under Options the window Compare Main Effects which offers: LSD (Least Significant Difference), Bonferroni and Sidak. We opt for Bonferroni and show the results in Table 8.20.

Table 8.20. Post hoc comparions carried out on the within-subject factor 'vowels' of Table 8.18; output of the SPSS module GLM, Repeated Measures

(I) VOWEL	VOWEL (J)	Mean Diff. (I–J)	Std. Error	Sig.	95% Conf. Interval for Difference Lower Bound	Upper Bound
1	2 ①	−11.750	1.902	.000	−17.330	−6.170
	3	1.500	2.303	1.000	−5.257	8.257
2	1 ①	11.750	1.902	.000	6.170	17.330
	3 ②	13.250	2.529	.002	5.832	20.668
3	1	−1.500	2.303	1.000	−8.257	5.257
	2	−13.250	2.529	.002	−20.668	−5.832

The output in Table 8.20 (comparable to the output for post hoc comparisons in one-way and multifactorial designs) indicates the following:

① A lot of redundant information is presented: The differences between the vowels 1 (/i/) and 2 (/a/) on the one hand and 2 and 1 on the other are of course the same, apart from the sign; they differ significantly.

② The same holds for the vowels 2 (/a/), and 3 (/I/). Thus the vowels /i/ and /I/ constitute a homogeneous subset (short vowels) and /a/ the subset of long vowels.

8.11. A split-plot design: within and between-subject factors

In the following example we show the analysis of data obtained in a split-plot design. This is a design with one or more between-subject factors and one or more within-subject factors. In research on speech and language behaviour the between-subject factors are often background factors like 'language/dialect', 'gender', 'SES' etc. In our example we compare the behaviour of two groups of stutterers. Five participants were randomly assigned to a treatment which focusses on articulatory competence (Group 1), whereas five other participants were assigned to a treatment focussing on emotional coping (Group 2). The 10 stutterers were measured twice on a scale which is supposed to measure articulation competence. The data are given in Table 8.21

The data matrix for the data of Table 8.21 is given in Table 8.22. The matrix shows which variables are required.

Table 8.22 again shows that each case, here the stutterer, has a value for each variable. Each case is marked for the group to which it belongs, group 1 or group 2. The variables T1 and T2 represent the two measures on ar-

Table 8.21. Scores of articulatory competence obtained in two groups of stutterers at two time points

		T1	T2
group 1	S_1	70.0	80.0
	S_2	60.0	60.0
	S_3	45.0	55.0
	S_4	65.0	70.0
	S_5	55.0	65.0
group 2	S_6	65.0	65.0
	S_7	55.0	60.0
	S_8	40.0	42.0
	S_9	38.0	35.0
	S_{10}	40.0	42.0

Table 8.22. The SPSS data matrix representing data set (a) from Table 8.21

group	T1	T2
1	70	80
1	60	60
1	45	55
1	65	70
1	55	65
2	65	65
2	55	60
2	40	42
2	38	35
2	40	42

ticulation competence. The syntax for this repeated measures design plus a between-subject factor is:

```
GLM
  T1 T2 BY group
  /WSFACTOR = time(2)
  /METHOD = SSTYPE(3)
  /CRITERIA = ALPHA(.05)
  /WSDESIGN = time
  /DESIGN = group .
```

The relevant SPSS output consists of two parts (Tables 8.23a and 8.23b):

① Table 8.23a deals with the within-subject factors and interactions with those factors. In our case the 'time' factor and the interaction 'time × group'. As the within-subject factor only comprises two levels, corrections for violations of the sphericity assumption are not relevant (see Section 8.3); that is why both Greenhouse-Geisser and the Lower Bound are left out. For illustrative purposes we only show the Huynh-Feldt correction. In our example both the factor 'time' and the interaction 'time × group' are significant. Group 1 appears to have made more progress in articulatory skills, presumably as a result of the articulation based therapy.

② Table 8.23b gives the results for the between-subject factor, in our case 'group'. The 'group' factor in itself is not significant ($p = 0.073$). As a matter of fact, the significance of a main factor is less relevant in light of the interaction of that factor with another factor, in our example: 'time'.

Table 8.23a. Part of the SPSS output of the GLM, Repeated Measures procedure for the two data sets of Table 8.21

Source		Type III Sum of Squares	df	Mean Square	F	Sig.
time ①	Sphericity Assumed	84.050	1	84.050	11.714	.009
	Huynh-Feldt	84.050	1.000	84.050	11.714	.009
time * group ②	Sphericity Assumed	42.050	1	42.050	5.861	.042
	Huynh-Feldt	42.050	1.000	42.050	5.861	.042
Error(time)	Sphericity Assumed	57.400	8	7.175		
	Huynh-Feldt	57.400	8.000	7.175		

Table 8.23b. Part of the SPSS output of the GLM, Repeated Measures procedure for the two data sets of Table 8.21

Source	Type III Sum of Squares	df	Mean Square	F	Sig.	Partial Eta Squared
Intercept	61272.450	1	61272.450	255.488	.000	.970
group ②	1022.450	1	1022.450	4.263	.073	.348
Error	1918.600	8	239.825			

8.12. Missing data

Missing data constitute a serious problem, especially in the analysis of observations obtained in repeated measures designs, where missing data occur in data obtained with subjects who participated in the experiment, but did not always yield an observation. This happens quite frequently in psycholinguistic experiments in which Reaction Times are measured. Sometimes more than 10% of the data is missing. Reasons are the distraction of participants, or participants who react too soon and thus do not yield valid data. There are a number of ways available to handle missing data in repeated measures designs. For reviews and details of ways of handling missing data we refer to Little and Schenker (1995), Little and Rubin (2002) and Allison (2002) and to Section 9 of Chapter 9. For illustrative purposes we present a simple randomized block design, with five participants measured three times, at t_1, t_2, and t_3. In data set A in Table 8.24 we present the complete data set, in data set B the set with two missing data: in subject 2 at t_2 and in subject 4 at t_3.

Table 8.24. Two data sets with five participants measured at t_1, t_2 and t_3; in data set B two observations are missing

	data set A			data set B		
	t_1	t_2	t_3	t_1	t_2	t_3
*s*1	3	4	4	3	4	4
*s*2	2	2	3	2	-	3
*s*3	5	5	6	5	5	6
*s*4	6	7	7	6	7	-
*s*5	4	5	5	4	5	5

The F value associated with the complete data is: $F_{2.8} = 12.667$, $p = 0.003$. What can we do if there are missing data as in data set B in Table 8.24? If we carry out the standard GLM, Repeated Measures procedure of SPSS cases 2 and 4, with the missing data, are skipped ('listwise deletion'). The resulting $F_{2.4} = 7.000$, $p = 0.049$. The deletion of two cases resulted in quite a high loss of power and consequently a much higher level of p. There are a number of alternatives to deal with the missing data; details are given in Chapter 9.

8.13. Suggestions for statistical reporting

EXAMPLE 1

The Reaction Times in our within-subject design were analyzed with the GLM Repeated measures procedure of SPSS. There were two within-subject factors: 'word type' (3 levels, fixed factor) and 'words' (10 levels, nested within 'word type', random). Thirty participants took part in the experiment. Twelve per cent of the data had to be discarded as they did not meet the criteria mentioned earlier. Both F1 and F2 were significant at the 5% level for the factor 'word type: $F1_{2,58} = 4.912$ (*p* value Huynh-Feldt corrected), MSE = 193425; $F2_{2,27} = 3.804$, MSE = 18304.4; homogeneity of variances was demonstrated by Levene's test. The effect size for F1: $\eta^2_{partial} = 0.629$, for F2 $\eta^2_{partial} = 0.220$.

EXAMPLE 2

The scores on the 'surprise' scale were analysed by analysis of variance (GLM repeated measures procedure of SPSS). There was one between-subject factor: 'Language background' (fixed) with two levels: English and Dutch, and one within-subject factor: 'Peak height of the fundamental frequency' (fixed), with five levels. The 'Language background' factor was significant at the 5% level: $F_{1,38} = 7.80$ ($p = 0.008$), MSE = 716.40, $\eta^2_{\text{partial}} = 0.170$. The ordinal interaction between 'Language background' and 'Peak height' was significant at the 5% level: $F_{4,38} = 11.20$ ($p = 0.000$, Huynh-Feldt adjusted), MSE = 716.40, $\eta^2_{\text{partial}} = 0.722$. The within-subject factor 'Peak height' was also significant: $F_{4,38} = 23.14$ ($p = 0.000$, Huynh-Feldt adjusted), MSE = 716.40, $\eta^2_{\text{partial}} = 0.708$.

8.14. Terms and concepts

- *F1*: An *F* ratio obtained by pooling over 'items' ('words') in a hierarchical design with for instance 'type of words' and 'items' as within-subject factors.
- *F2*: An *F* ratio obtained by pooling over 'subjects' ('participants') in a hierarchical design with for instance 'type of words' and 'items' as within-subject factors.
- *Multivariate analysis of variance (MANOVA)*: Analysis of variance carried out on more than one dependent variable.

- *Quasi F ratio*: An approximation of the *F* statistic, obtained by summing mean squares to form appropriate numerators and denominators for the *F* ratio.
- *Randomized block design*: A design characterized by grouping participants into blocks in such a way that units within a block are as similar as possible with respect to the dependent variable, while the units in different blocks are less similar.
- *Repeated measures design*: A design in which participants are measured in more than one condition, also called a within-subject design.
- *Split-plot design*: A design which is a combination of a completely randomized factorial design and a repeated measures design (or randomized block design). The repeated measures are restricted to a subset of the factors, whereas the participants are distinguished in separate groups by at least one factor.
- *Within-subject factor*: A factor which comprises different conditions in which the same subject is measured.

8.15. Exercises

1. A psycholinguist investigates the reaction times of listeners asked to detect visually presented targets, which can be a CV or CVC sequence. The targets occur in two types of words, which share the initial CVC, but differ in the position of the syllable boundary (French: 'balcon' vs. 'balance'). Each type of word has 10 tokens. Each of the listeners has to react to the same stimuli. Give the design of this experiment. Are the factors crossed or nested?
2. Suppose the following reaction times (measured in milliseconds) to four words (W1, W2, W3, W4) are obtained. The words are nested within the 'stress' factor. Five participants participated in the experiment.

	− stress		+ stress	
	W1	W2	W3	W4
subject 1	500	480	480	465
subject 2	480	460	440	430
subject 3	490	485	420	400
subject 4	520	550	490	480
subject 5	530	510	495	490

Use a computer program to carry out an analysis of variance on these data. Determine $F1, F2, F1 \times F2, F'$, and minF' for the factor 'stress'; both participants and words constitute random factors.

3. Are Huynh-Feldt corrected *df* values expected to yield smaller or larger *p* values?
4. Should a Huynh-Feldt correction be applied to the degrees of freedom associated with the between-subject or within-subject factors?
5. Assume we obtained the following *F* ratio: $F_{5.20} = 2.83$. H-F's ε is 0.8. Is the H-F corrected *F* ratio significant at the 5% level?
5. A researcher investigates the reaction of newborn babies to four types of consonant-vowel (CV) sequences. Ten tokens of the four CV sequences are presented to each of the four babies. The dependent variable is 'sucking behaviour', expressed in the number of sucks. Determine the error terms for testing the factor 'CV-type' and 'babies', when the former is regarded as fixed and the latter as random.
6. Is it possible to test all sources of variation in the preceding experiment when only one sample of each CV-sequence is presented to the babies?

Chapter 9
Alternative estimation procedures and missing data

9.1. Preview

In this chapter we discuss an alternative estimation technique used in analysis of variance: *(maximum) likelihood* (ML) estimation and a problem which we already discussed in preceding chapters, that of *missing data*. Likelihood estimation in analysis of variance has become available in most statistical packages, and is considered a good alternative to standard ANOVA in case of missing values. In Section 9.2 the concept of likelihood estimation is introduced. In the next sections, i.e. 9.3 and 9.4, the use of likelihood estimation in analysis of variance is discussed and illustrated on the basis of a number of examples and SPSS output. Finally, we discuss problems associated with missing data and give solutions in Section 9.5.

9.2. Likelihood estimation

Analysis of variance, as presented in the preceding chapters, is not as elegant in theory and practice as it may seem at first sight. There are a number of negative points:

1. The estimators of variance components like σ_α^2 can turn out to be negative, which is obviously incorrect (an 'embarrassment', as it is called by Searle, Casella, and McCulloch 1992). Such a negative estimator can result from a specific combination of – in the case of a one-way analysis – values of MS_{between} and MS_{within}. These MS values are used to estimate the values of σ^2s on the basis of their expected values: $E(MS_{\text{within}}) = \sigma_\varepsilon^2$ and $E(MS_{\text{between}}) = n\sigma_\alpha^2 + \sigma_\varepsilon^2$.
2. In most designs the presence of missing data requires the use of quite complicated and sometimes fortuitous procedures (see Section 7.9 for the four different types of sums of squares calculated on the basis of regression analysis) or the loss of data, with associated loss of power.

An important development in statistics (already suggested by Fisher in 1922 for analysis of variance), but only implemented in most software packages in the late 1990s is that of maximum likelihood (ML) estimation. This estimation consists of finding the parameter θ of the population which maximizes the likelihood of observing data x. An example is the proportion of successes, θ in a population, and the proportion of observations of successes in the sample, x. Maximum likelihood estimators are obtained on the basis of likelihood functions. These are functions which express the likelihood to observe given data on the basis of specific parameter values. The general expression for the likelihood function is $L(x|\theta)$. In fact it is the probability distribution of x, given θ. A simple likelihood function is the one for binomial data. Let us assume that we found 3 'successes' in a sample of 10 observations; we have to estimate the value of θ, this is π for binomial data, which maximizes the likelihood of observing the percentage of successes. We try three values of π: 0.20, 0.30 and 0.40 by putting them into the associated likelihood function L:

$$L = \frac{n!}{k!(n-k)!} \times \pi^k \times (1-\pi)^{n-k} \qquad (1)$$

with n being the number of observations and k the number of 'successes'. This is the well-known formula to calculate the probability of observing k successes when the parameter of the nominal distribution is π. The values of L for the three values of π are: 0.2013, 0.2668 and 0.2150 respectively. Obviously we opt for $\pi = 0.30$, as it results in the maximum value of L: 0.2668. We found this result by simply trying a number of values of π; in practice we need a general method to find the value of θ which maximizes the likelihood to observe k successes. To that end we have to calculate derivatives of the likelihood function and set them at zero (a well-known technique of basic calculus); they yield the value(s) of θ(s) for which the likelihood function attains its maximum or minimum value. Second derivatives make clear whether specific values yield a maximum or a minimum. Most likelihood functions are quite complicated, and many of them are not available in closed form, which means that they cannot be solved on the basis of equations. If no closed form exists, the solution can only be found by using iterative procedures implemented in computer packages. In order to show the degree of complexity for relatively simple problems, we give an example for a familiar result: the ML estimator of μ on the basis of a sample consisting of $x_1, x_2, \ldots.. x_n$. Let us assume that x is $N(\mu, \sigma^2)$ distributed, and that σ^2 is known. The probability of observing a value x_i is governed by the well-known probability density function (PDF) given in (2):

$$f(x;\mu,\sigma^2) = \frac{1}{\sqrt{2\pi\sigma}} e^{\frac{-(x_i-\mu)^2}{2\sigma^2}} \qquad (2)$$

The likelihood function is the joint distribution of all independent values of x; thus we have to calculate the product of the above PDF for all values of x_i.

$$L(x_1, x_2, ..x_n|\mu,\sigma^2) = \frac{1}{\sqrt{2\pi\sigma}} e^{\frac{-(x_1-\mu)^2}{2\sigma^2}} \times .... \times \frac{1}{\sqrt{2\pi\sigma}} e^{\frac{-(x_n-\mu)^2}{2\sigma^2}} \qquad (3)$$

In order to estimate the value of μ, we have to maximize L. To that end we first start a simplifying operation, which yields:

$$L = (\frac{1}{2\pi\sigma^2})^{\frac{n}{2}} e^{-\frac{1}{2\sigma^2}\sum_i (x_i-\mu)^2} \qquad (4)$$

In this formula $\frac{n}{2}$ and $\sum$ reflect the original multiplication operation. When one takes logarithms, the calculations become simplified as products turn into additions ($\log(a \times b) = \log(a) + \log(b)$). That is why likelihood functions are transformed into logarithmic expressions. The natural logarithm ln (with base e which is 2.71828) is used as the likelihood function involves e, and $\ln(e) = 1$. The logarithmic transformation turns equation (4) into (5):

$$\ln L = -\frac{n}{2}\ln 2\pi - \frac{n}{2}\ln\sigma^2 - \frac{1}{2\sigma^2}\sum_i (x_i - \mu)^2 \qquad (5)$$

Taking the partial derivative with respect to μ we get:

$$\frac{\delta \ln L}{\delta \mu} = 1/\sigma^2 \sum_i (x_i - \mu). \qquad (6)$$

We obtain the ML estimator for μ by setting this derivative to zero; thus we get the following estimator of μ: $\sum x_i/n = \bar{x}$, a familiar result.

One may be inclined to think that ML estimators are the same as other more familiar estimators, such as Ordinary Least Squares (OLS). This is not true. In analysis of variance ML estimators can be quite different from OLS estimators.

9.3. Likelihood estimation in analysis of variance

As we said above, likelihood functions for analysis of variance are quite complicated. We give L for a one-way analysis, with μ, σ_α^2 and σ_ε^2 as parameters:

$$L = \frac{e^{[-\frac{1}{2\sigma_\varepsilon^2}[\sum_j \sum_i (x_{ij}-\mu)^2 - \sum_j \frac{\sigma_\alpha^2}{\sigma_\varepsilon^2 + n_j \sigma_\alpha^2}(x_{.j} - n_j \mu)^2]]}}{(2\pi)^{1/2N} \sigma_\varepsilon^{2[1/2(N-k)]} \prod_j (\sigma_\varepsilon^2 + n_j \sigma_\alpha^2)^{1/2}} \quad (7)$$

There is a variant of Maximum Likelihood Estimation, called Restricted Maximum Likelihood Estimation (REML), also available in SPSS. The difference from ML is that it only maximizes that part of L which does not contain μ. To put it differently: Only those variance components are estimatord which are free from fixed effects. The population parameter μ is a fixed effect (labelled 'intercept' in SPSS and other statistical packages). This leaves us with three estimators of effects:

- the conventional ANOVA estimators, also called Ordinary Least Squares (OLS) estimators
- Maximum Likelihood (ML) estimators
- Restricted Maximum Likelihood estimators (REML).

In SPSS the Mixed Models procedure yields the last two estimators and the procedure GLM, Generalized Linear Models the first type of estimators.

In Table 9.1 we list the ML and ANOVA estimators for two of the three relevant parameters of a one-way design: $\hat{\sigma}_\alpha^2$ (the variance of the group effects in the population for a random effect design) and $\hat{\alpha}_k$ for a fixed design, and the random error $\hat{\sigma}_\varepsilon^2$.

The estimators differ as a function of two aspects: The grouping variable is either a fixed or a random factor, and the presence or absence of missing data: *unbalanced* vs. *balanced* design. For unbalanced designs, no closed forms exist (see McCullooch & Searle 2001). In order to find estimators with which this equation can be solved, iterative procedures have to be used. If a solution cannot be found, a message like the following will be issued: CONVERGENCE CRITERIA NOT SATISFIED.

In the Mixed Models procedure for ML estimators in SPSS the presence of a fixed effect in balanced designs leads to a different value of MS_E, and consequently to a different estimator of σ_ε^2 as restrictions in the degrees of freedom used in estimating the αs are not taken into account. MS_E is obtained by dividing SS_E by $k \times n$, instead of $k \times (n-1)$. This also results in a different value of the F ratio used to test the effect of the 'group' factor, as we will see later. For *random* models in balanced designs the OLS estimators are the same as for the *fixed* models (this is only true for one-way designs).

Table 9.1. OLS and ML estimators in analysis of variance; k denotes number of levels of the one-way design, x_{ij} an observation i in group j

OLS	*balanced*	*unbalanced*
fixed	$SS_A = \sum_j (\bar{x}_{.j} - \bar{x}_{..})^2$ $SS_E = \sum_j \sum_i (x_{ij} - \bar{x}_{.j})^2$ $MS_A = SS_A/(k-1)$ $MS_E = SS_E/(k(n-1))$ $\hat{\alpha}_j = \bar{x}_{.k} - \bar{x}_{..}$ $\hat{\sigma}_\varepsilon^2 = \frac{\sum_j \sum_i (x_{ij} - \bar{x}_{.j})^2}{k \times (n-1)} = MS_E$ Test: $F = MS_A/MS_E$	$SS_A = \sum_j n_j (\bar{x}_{.j} - \bar{x}_{..})^2$ $SS_E = \sum_j^k \sum_i^{n_j} (x_{ij} - \bar{x}_{.j})^2$ $MS_A = SS_A/(k-1)$ $MS_E = SS_E/(k(n-1))$ $\hat{\alpha}_j = \bar{x}_{.k} - \bar{x}_{..}$ $\hat{\sigma}_\varepsilon^2 = \frac{SS_E}{N-k} = MS_E$ Test: $F = MS_A/MS_E$
random	$\hat{\sigma}_\alpha^2 = (MS_A - MS_E)/n$ for other terms, see under 'fixed' Test: $F = MS_A/MS_E$	$\hat{\sigma}_\alpha^2 = \frac{(MS_A - MS_E)}{(N - \sum_j n_j^2/N)/(k-1)}$ for other terms, see under 'fixed' Test: $F = MS_A/MS_E$
ML	*balanced*	*unbalanced*
fixed	SS_A, SS_E, MS_A like OLS $MS_E = SS_E/kn$ other terms like under 'fixed' and OLS Test: $F = MS_A/MS_E$	no closed form exists Test: Wald statistic $\hat{\sigma}_\alpha^2/SE_{\hat{\sigma}_\alpha^2}$
random	if $(1 - 1/k)MS_A \geq MS_E$ $\hat{\sigma}_\alpha^2 = \frac{(1-1/k)MS_A - MS_E}{N}$ if $MS_A < MS_E$ $\hat{\sigma}_\alpha^2 = 0$ if $(1 - 1/k)MS_A \geq MS_E$ $\hat{\sigma}_\varepsilon^2 = MS_E$ if $MS_A < MS_E$ $\hat{\sigma}_\varepsilon^2 = SS_T/(k \times n)$ Wald statistic $\hat{\sigma}_\alpha^2/SE_{\hat{\sigma}_\alpha^2}$	no closed form exists Wald statistic $\hat{\sigma}_\alpha^2/SE_{\hat{\sigma}_\alpha^2}$

Testing effects

When we carry out a standard analysis of variance (for instance a one-way analysis), we are used to testing effects on the basis of F ratios. The ML and REML approaches use different procedures:

- For random components both ML and REML use the Wald statistic, which involves the quotient of the estimator of the variance component and the Standard Error of that component: $\sigma_\alpha^2/SE(\sigma_\alpha^2)$. $SE(\sigma_\alpha^2) = ((2 \times$

$\sigma_\varepsilon^4)/n^2) \times [1 + n \times (\sigma_\alpha^2/\sigma_\varepsilon^2)^2/(a^2/(a-1)) + 1/(a \times (n-1))]$.

- The estimators of σ_α^2 by ML and REML differ, so do the *SE*s of these variance components. Consequently, the Wald statistic will differ as well.
- When the model involves fixed effects, ML and REML will yield different results, as REML takes the degrees of freedom used for estimating fixed effects into account. Paradoxically, it does not estimate these fixed effects.

The SPSS syntax for a one-way analysis with the MIXED procedure is as follows:

```
MIXED dep BY group
  /FIXED= group
  /METHOD=ML.
```

9.4. Two examples: a balanced and an unbalanced design

Example 1: balanced data

In Table 9.2 we reproduce Table 4.1, data set B, with a data set of three groups and three observations in each group. The relevant statistics obtained in the Compare Means, One Way ANOVA analysis of SPSS are: $SS_A = 24.00$, $df = 2$, $MS_A = 12.00$ and $SS_E = 6.00$, $df = 6$, with $MS_E = 1.00$.

Table 9.2. A hypothetical data set of regional accent scores for three age groups of three participants; reproduction of Table 4.1, data set B

	age group		
	I	II	III
	3	7	1
	4	6	2
	5	5	3
$\bar{X}$	4	6	2

We distinguish two situations: (1) 'group' as a fixed factor, and (2) 'group' as random factor. We present the results for a) the conventional ANOVA approach (SPSS Compare Means, OneWay ANOVA) and b) the ML and REML estimators provided by SPSS Mixed Models.

In the conventional ANOVA approach the F value associated with the fixed factor 'group' is $F_{2,6} = 12.000$ ($MS_{\text{group}}/MS_{\text{within}} = 12/1$), $p = 0.008$. For one-way models, it does not matter whether the 'group' factor is random or fixed; in both cases the same error term is used for the F ratio. In case the 'group' factor is random the estimator of $\sigma_\varepsilon^2 = MS_{\text{E}}$, and the estimator of σ^2_{group} ($= \hat{\sigma}^2_{\text{group}}$) can be obtained as follows (see Chapter 4):

$$\begin{aligned} E(MS_{\text{between}}) &= n\sigma^2_{\text{group}} + \sigma^2_\varepsilon \quad (8) \\ E(MS_{\text{within}}) &= \sigma^2_\varepsilon \\ E(MS_{\text{between}}) - E(MS_{\text{within}}) &= n\sigma^2_{\text{group}} \\ 12 - 1 &= 3\sigma^2_{\text{group}} \\ \hat{\sigma}^2_{\text{group}} &= 3.66 \end{aligned}$$

Table 9.3a. (Part of the) SPSS output from an analysis of variance Mixed Models procedure applied to the data set of Table 9.2; The 'group' factor is fixed

Type III Tests of Fixed Effects ①

Source	Numerator df	Denominator df	F	Sig.
Intercept	1	9.000	216.000	.000
GROUP	2	③ 9.000	18.000	.001

Table 9.3b. (Part of the) SPSS output from an analysis of variance (Mixed Models procedure) applied to data set in Table 9.2; estimators of covariance parameters, with 'group' as a fixed factor

Estimators of Covariance Parameters ①

Parameter	estimator	Std. Error
Residual ②	.6666667	.3142697

An analysis of variance was carried out on the data of Table 9.2 with the We analyzed the data in Table 9.2 with the Mixed Models procedure of SPSS. The results are given in Tables 9.3a to 9.3c:

① In the SPSS analysis with MIXED models, the distinction between random and fixed factors becomes important once again. If the factor is FIXED, the relevant output consists of two tables, one called 'Type III tests of Fixed Effects' (with the F ratio testing the fixed effect), and the other called 'Estimators of Covariance Parameters'.

② The table with covariance parameters gives the estimators of the random components, the only one being $\hat{\sigma}_{\varepsilon}^2$ in the case of a fixed model: 'Residual'.
The F ratio is larger (and consequently the power associated with the test as well) than the one obtained with the ANOVA model. The differences are the result of the fact that the error term is not the conventional MS_{error} (MS_{E} obtained in the ANOVA approach), in our example 1.00, but the ML estimator of the random component σ_{error}^2, which is estimated in the case of a fixed effect model by $SS_{\text{E}}/(k \times n)$: $6/9 = 0.667$.

③ The degrees of freedom associated with the error term is not $3 \times (3-1) = 6$, but 9, as in the ML estimation all data - here 9 - are used.

If the factor 'group' is a *random* factor, the test is $\sigma_{\text{group}}^2 = 0$, whereas for a fixed model, the test is: all $\alpha_k = 0$. In the former case, the SPSS procedure MIXED with ML-estimators will not deliver an F test for the factor 'group', but a Confidence Interval (= *ci*) for σ_{group}^2 and the Wald test for the hypothesis that $\sigma_{\text{group}}^2 = 0$. The relationship between the *ci* and the result of the test is not as obvious as it is with customary z and t tests. Here the *CI* is calculated for the hypothesis that σ^2 is zero. If there is reason to assume that σ^2 is not zero (as in the Wald test), the *ci* is not in line with the test anymore. Thus we recommend not to pay too much attention to the confidence intervals of the σ^2s.

In Tables 9.4a and 9.4b we show the estimators of both variance components and the associated Wald tests as given by SPSS, with random factor 'group'. Under the heading Estimates of Covariance Parameters (see ① under Table 9.3b) we find estimators of both σ_{group}^2 and σ_{ε}^2. This heading is used as for the random model the covariance between observation x_i in group j and observation in group $x_{i'}$ in the same group j can be shown to be:

$$\begin{aligned} \text{cov}(x_{ij}, x_{i'j}) &= \text{cov}(\mu + \alpha_j + \varepsilon_{ij}, \mu + \alpha_j + \varepsilon_{i'j}) \quad (9) \\ &= \sigma_{\text{group}}^2 \quad \text{for } i \neq i' \end{aligned}$$

Using the formula for the ML estimator of $\sigma_{\text{group}}^2 = [(1 - 1/k)MS_{\text{A}} - MS_{\text{E}}]/n$ we get: $[((1 - 1/3)12 - 1)]/3 = 2.33$, the same value as given by SPSS.

Table 9.4a. (Part of the) SPSS output from an analysis of variance (procedure Mixed Models) applied to data set in Table 9.2; estimates of covariance parameters with 'group' as a random factor

Estimates of Covariance Parameters ①

Parameter		Estimate	Std. Error	Wald Z	Sig.
Residual		1.0000000	.5773503	1.732	.083
GROUP	Variance	2.3333333	2.1858128	1.067	.286

Table 9.4b. (Part of the) SPSS output from an analysis of variance (procedure Mixed Models) applied to data set in Table 9.2; confidence interval, with 'group' as a random factor

Parameter		95% Confidence Interval	
		Lower bound	Upper Bound
Residual		.3225214	3.1005693
GROUP	Variance	.3720410	14.6339901

The SPSS syntax for the output presented in Tables 9.3a and 9.3b is as follows:

```
MIXED age BY group
  /METHOD=ML
  /PRINT = TESTCOV
  /FIXED= group.
```

The last line defines 'group' as a fixed effect. The results for 'group' as a random factor were given in Tables 9.4a and 9.4b. These results can be obtained by the following syntax:

```
MIXED age BY group
  /METHOD=ML
  /PRINT = TESTCOV
  /RANDOM= group.
```

Example 2: Unbalanced data

For the analysis of unbalanced data in a one-way design three approaches are available with different results: ML, REML and ANOVA. In Table 9.5, we give the estimators of variance components and the F and Wald tests for the three estimation procedures for 'group' as fixed and as random factor, and one missing data (case 3 in group 3) in the data set of Table 9.2.

Table 9.5. Comparison of different approaches to an unbalanced design (Table 9.2, with the last observation of age group III set as 'missing'

	ANOVA	REML	ML
Fixed	$F_{2,5} = 13.452$ $p = .010$ $\hat{\sigma}^2_\varepsilon = 0.900$	$F_{2,5} = 13.452$ $p = .010$ $\hat{\sigma}^2_\varepsilon = 0.900$	$F_{2,8} = 21.667$ $p = .001$ $\hat{\sigma}^2_\varepsilon = 0.562$
Random	$F_{2,5} = 13.452$ $p = .010$ $\hat{\sigma}^2_\alpha = 4.300$ $\hat{\sigma}^2_\varepsilon = 0.900$	Wald = 4.6719 $p = .356$ $\hat{\sigma}^2_\alpha = 4.672$ $\hat{\sigma}^2_\varepsilon = 0.902$	Wald = 1.080 $p = .280$ $\hat{\sigma}^2_\alpha = 2.976$ $\hat{\sigma}^2_\varepsilon = 0.905$

For 'group' as fixed effect both the ANOVA and REML approaches yield the same results; the ML approach appears to have a greater power. For 'group' as a random factor and an unbalanced design the results are quite heterogeneous. Westfall (1987) showed that for the simultaneous estimation of $\hat{\sigma}^2_\alpha$ and $\hat{\sigma}^2_\varepsilon$ ML estimation is the better choice.

Goodness of fit statistics

SPSS also gives values of goodness of fit statistics; most of them are equivalent alternatives of the -2 Log Likelihood ($-2\ln L$) statistic, which can be easily derived from equation (5). It is:

$$\ln L = -\frac{n}{2}\ln 2\pi - \frac{n}{2}\ln\sigma^2 - \frac{1}{2\sigma^2}\sum_i^n (x_i - \mu)^2 \qquad (10)$$

When we multiply both sides by -2, we obtain the following goodness of fit index:

$$-2\ln L = n\ln 2\pi + n\ln\sigma^2 + \frac{1}{2\sigma^2}\sum_i (x_i - \mu)^2 \qquad (11)$$

The last term is equal to n, thus we obtain:

$$-2\ln L = n\ln 2\pi + n\ln\sigma^2 + n \qquad (12)$$

The interesting part of this formula is the middle one on the right, which involves an estimator of σ^2. This estimator varies as a function of the number of parameters used to predict the observations. The other components of the equation involve n and π, and reflect the origin of the equation: the likelihood of observing the sampled data (cf. equation (5)). There are quite a number of goodness of fit indices. For our data (if the 'group' factor is fixed) we give the calculations needed to obtain -2 ln likelihood (called -2 Log Likelihood in SPSS) and Akaike's Information Criterion (AIC), on the basis of $N = 9$ and the estimator of $\sigma_{\varepsilon}^2 = 0.667$). -2 ln L $= 9 \times \ln(2 \times 3.141592) + 9 \times \ln(0.667) + 9 = 21.89171$.

Akaike's Information Criterion (AIC) contains a penalty for the number of parameters used to predict the observations. In our case there are four parameters: the overall mean and the three means of the groups: AIC = -2(ln likelihood - number of parameters); this results in $21.89171 + 2\times$ the number of parameters (4) = 29.89171. The smaller these coefficients are, the better the fit of the model is.

If the 'group' factor is random, we obtain different values for the goodness of fit indices, as the estimator of σ_{ε}^2 is different (1.000).

Table 9.6. (Part of the) SPSS output from an analysis of variance (procedure MIXED) applied to the data set in Table 9.2; information criteria for 'group' as a fixed factor

-2 Log Likelihood	21.892
Akaike's Information Criterion (AIC)	29.892
Hurvich and Tsai's Criterion (AICC)	39.892
Bozdogan's Criterion (CAIC)	34.681
Schwarz's Bayesian Criterion (BIC)	30.681

Of course, these indices are only useful when different numbers of parameters can be used to predict the observations. That is why we present a data set in which we could either use one factor (3+1 = 4 parameters, corresponding with 3 means of the 'treatment' factor plus the overall mean) or two factors (3 x 2 + 1 = 7 parameters, corresponding to 3 means of the 'treatment' factor, 2 of the 'education' factor and the overall mean) to represent the observations. We refer to Table 5.2 of Chapter 5 which is reproduced in Table 9.7.

We have two options: either treating the obtained data in a one-factorial or in a two-factorial design. Using the SPSS Mixed Models option with ML as approach for the estimation of the variance components we got the following

Table 9.7. Two-factorial design (fixed factors) for the data of Table 5.2

	treatment		
education	I	II	III
high school	1	3	9
	2	4	10
university	4	8	14
	5	9	15

estimators of σ_{error} and the goodness of fit indices '-2 Log Likelihood' and AIC:

- Two factors in the model (equivalent to $3 \times 2 + 1 = 7$ parameters): σ^2_{error} = 0.500; -2 Log Likelihood= 17.027, AIC = 19.027; these values can be found in Table 9.8a.
- One factor in the model ('age', equivalent to 3 + 1 = 4 parameters): σ^2_{error} = 6.888; -2 Log Likelihood = 47.069, AIC = 49.069 (SPSS table not shown here).

In this case adding one factor to the model (viz. 'education' with two levels) clearly has a positive effect on the values of the Information Indices. The information criteria are displayed in smaller-is-better forms, as SPSS reminds us.

Table 9.8a. (Part of the) SPSS output for the two-factor model with Information criteria (= goodness of fit indices)

-2 Log Likelihood	17.027
Akaike's Information Criterion (AIC)	19.027

In Table 9.8b we present the results of the MIXED procedure when both 'education' and 'treatment' are included in the model.

- A proportion of 0.60 is observed in a sample. What is the maximum likelihood estimator of the associated π?
- What is an OLS estimator?
- A one-way analysis of variance is carried out. What is the difference between the associated tests when the independent variable 'group' is a fixed or a random factor?
- Will the –2 Log Likelihood goodness of fit statistic increase or decrease when the number of parameters used increases?
- What is the difference between the AIC and the –2 likelihood index?

Table 9.8b. (Part of the) SPSS output for the two-factor model with ANOVA results for the data of Table 9.7

			Unique Method				
			Sum of Squares	df	Mean Square	F	Sig.
Dep	Main Effects	(Combined)	224.333	3	74.778	149.556	.000
		ED	56.333	1	56.333	112.667	.000
		TREAT	168.000	2	84.000	168.000	.000
	2-Way interactions	ED * TREAT	2.667	2	1.333	2.667	.148
	Model		227.000	5	45.400	90.800	.000
	Residual		3.000	6	.500		
	Total		230.000	11	20.909		

9.5. Imputation procedures for missing data

Missing data constitute a serious problem, especially in the analysis of observations obtained in repeated measures designs. The absence of equal numbers of observations can be due to a sampling problem (some subpopulations may yield fewer samples than others), non-response (people do not fill in their answer forms or refuse to participate) or even technical problems (a computer may not have recorded some of the responses). All these situations result in unequal numbers of participants and thus in unequal numbers of observations. In Chapter 7, Section 5, we discussed so-called unbalanced designs for factorial designs. Solutions were offered by the introduction of regression analysis. The regression analysis called for different approaches to enter the independent variables in the equations. For the analysis of repeated measures designs with missing data these approaches are not available. That is why we discuss, once again, missing values and ways of dealing with them.

Rubin (1976) distinguishes four different mechanisms which may underlie the occurrence of missing data and which may guide the use of specific procedures to deal with missing data:

- MCAR = Missing Completely At Random: The probability of data X missing is not dependent on the value of that variable, nor on any other variables in the sample. If, for instance, data are more likely to be missing in a certain age group, the missings are not MCAR.
- MAR = Missing At Random: The probability of data X missing is *not* dependent on the value of X, if we control for other variables in the analysis. When, for instance, the willingness of participants to report

the use of dialect at home (the dependent variable is 'use') appears to be dependent on the variable 'age group', but within the age group 'age' itself does not determine the probability of the occurrence of missing data, the data are said to be MAR. If the probability of missing data is still dependent on age, the missings are not MAR.

- IGNORABLE: The missings are MAR and the parameters which determine the occurrence of missing data are unrelated to the estimated parameters (= factors in ANOVA). Allison (2002) treats MAR and IGNORABLE as equivalent conditions.
- NONIGNORABLE: If data are neither MAR nor IGNORABLE, they are labelled NONIGNORABLE. In that situation appropriate modelling of the response mechanism is required, which is not always a simple task.

Missing data in completely randomized factorial designs can either be handled with the procedures developed for unbalanced designs, which were discussed in Chapter 7, or by the use of ANOVA procedures based on Maximum Likelihood estimation. Most statistical packages, however, do not have a default procedure to process missing data in repeated measures designs, and just skip the cases with one or more missing data (the so-called *listwise deletion*). Especially in experiments with many missing data this may result in a serious reduction of available cases and therefore in an undesirable reduction of power. If participants with a specific background yield more missings than other participants, the validity of the statistical inferences may be questioned.

In repeated measures designs missing data occur in data vectors of subjects who participated in the experiment, but did not always yield an observation. This happens quite frequently in psycholinguistic (and other) experiments in which reaction times are measured; sometimes more than 10% of the data is missing. Common reasons are distraction of participants, or participants who react too soon and thus do not yield valid data. There are a number of ways available to deal with missing data in repeated measures designs. For reviews and details of handling missing data we refer to Little and Schenker (1995), Little and Rubin (2002) and Allison (2002).
Possible solutions are:

- Adjust degrees of freedom; this procedure is quite cumbersome and is not often applied (see for details Glantz and Slinker(1990) and Milliken and Johnson (1997).
- Resort to a procedure which is less sensitive to missing data, like Multi-

level Analysis (see Chapter 10, Section 4).

- Pool over participants and/or levels of the within-subject factor (see Chapter 8, Section 5) in a repeated measures design and calculate F1 and F2 respectively. The pooling procedure obscures the presence of missing data and F2 will ignore the within-participant covariances.
- Carry out an imputation procedure; imputation procedures fill in missing data on the basis of the available data. There are quite a number of imputation procedures, some of which are integrated into the current software packages, while others need dedicated software.

Below we review some popular imputation procedures.

Assume we have a simple randomized block design with five participants measured three times, at t_1, t_2, and t_3. In Table 9.9 we present the complete data set A and data set B with two missing data: in subject 2 at t_2 and in subject 4 at t_3.

Table 9.9. Two data sets with five subjects measured at t_1, t_2 and t_3; in Table 9.9b two observations are missing

	data set A			data set B		
	t_1	t_2	t_3	t_1	t_2	t_3
s_1	3	4	4	3	4	4
s_2	2	2	3	2	-	3
s_3	5	5	6	5	5	6
s_4	6	7	7	6	7	-
s_5	4	5	5	4	5	5

In the following we describe a number of alternatives. We will not report the Huynh-Feldt corrected p values, but the p values based on the nominal degrees of freedom minus 2 (2 = number of imputed values) in case of imputations: $df_2 = 6$. The F value associated with the complete data is: $F_{2,8} = 12.667$, $p = 0.003$.

What are the possibilities? If we carry out a standard GLM, repeated measures procedure of SPSS cases 2 and 4, which contain the missing data, are skipped ('listwise deletion'). The resulting $F_{2,4} = 7.000$, $p = 0.049$. The deletion of two cases resulted in quite a high loss of power and consequently a much higher level of p.

The *first* imputation procedure is to fill in the *mean value of all observed data*, in our example 4.538 and to reduce the df, in our case by 2 (two missing values are imputed on the basis of the present data). Obviously this procedure will generally result in an underestimation of the error, as the variation of the data is smoothed. In our example, however, the associated $F_{2,6} = 2.337$, $p = 0.178$. Apparently filling in the missing values with one and the same value equalizes the effects of the factor at issue in our example.

The *second* imputation procedure is to fill in the *means of the data obtained at* t_2 *and* t_3, which are 5.25 and 4.50 respectively. Thus we obtain $F_{2,6} = 2.346$, $p = 0.177$ (calculated with the SPSS function 1-CDF.F(q,df1,df2). Just filling in the mean of the values obtained at t_2 and t_3 will not do justice to the specific parameter associated with the subjects at issue.

The *third* imputation procedure is the one suggested by Allan and Wishart (1930). In their formula the estimator of a missing value in treatment T and block (participant) B of a randomized block design with B blocks and T treatments is:

$$E(x_{b,t}) = \frac{Tx_{+}^{t} + Bx_{+}^{b} - x_{+}}{(T-1)(B-1)} \qquad (13)$$

in which x_{+}^{t} is the sum of observed values for treatment t, x_{+}^{b} the sum of observed values for block b, and x_{+} the sum of all observed values. In our case the imputed values are 3.625 for $x_{2,2}$ and 7.50 for $x_{4,3}$. The resulting $F_{2,6}$ is 13.746, $p = 0.006$, close to the value obtained with the complete data set.

The *fourth* imputation procedure is called the *hot deck* procedure; it consists of filling in the missing data by values randomly drawn (with replacement) from similar responding cases. The draw can be carried out with and without replacement. In many designs it is not always obvious how these values should be drawn, because we do not know what these cases are. In our example the moments in time are not similar, nor are the participants. In designs with a hierarchical structure, like the one presented in Chapter 8 it is more obvious to find similar cases, as the words nested within 'type' can often be assumed to be thus. A random draw of the RT to a word of one type may be a good imputation.

The *fifth* procedure is the *regression approach.* This imputation procedure uses regression analysis carried out on the variable with the missing(s) as criterion variable and the variables without missings as predictors. The resulting regression weights (B_0, B_1, etc.) are used to predict the missing value(s). The dataset in Table 9.10 with two variables, V_1 and V_2, is used as an example.

The regression equation is $V1 = 16.637 + 0.760V2$. For case 5 we get $16.637 + 0.760 \times 60 = 62.237$ and for case 7 $16.637 + 0.760 \times 40 = 47.036$.

Table 9.10. Bivariate data, with observations on variables V_1 and V_2

V_1	V_2
80.0	70.0
70.0	50.0
60.0	48.0
55.0	45.0
–	60.0
50.0	55.0
–	40.0
70.0	80.0
45.0	55.0
30.0	27.0

As an option the residuals of randomly selected cases can be added to the imputed data. With standardized residuals added to the predicted values we obtain 63.588 and 46.194 as imputed values for case 5 and case 7 respectively. The addition of residuals corrects the reduction of the standard errors which are the result of the imputation procedure. Using the regression approach of the MVA procedure of SPSS for the repeated meadures data of Table 9.9 we obtain 5.250 as imputed value for cell s_2t_2 and 5.833 for s_4t_3 (residuals not added).

The following steps in the Point-And-Click (PAC) window of SPSS lead to the imputed values for the two missing values of V_1 in Table 9.10:

window	Data View	
	click on	Analyze
	go to	Missing Value Analysis
	click on	'V1' and 'V2'
	insert	under Quantitative Analysis
	click on	Regression
	click on	Regression (third button)
window	Estimation Adjustment	
	click on	Residuals
	click on	Save completed data
	click on	File (give a name to file to be saved)
	click on	Continue
	click on	OK

The *sixth* imputation procedure is called EM imputation. EM stands for Expectation Maximization. The Expectation part is the one in which parameters are estimated of the distribution(s) from which the data are sampled; in the Maximization part data are filled in on the basis of the estimated parameters. The steps are carried out by iteration until convergence (imputated data do not change much) is achieved:

- Estimate parameters of the distribution in the population.
- Reestimate the missing values assuming the estimated parameters are correct.
- Reestimate the parameters until convergence is reached.

Discussing the ins and outs of EM-algorithms is beyond the scope of our book. In the following simple example with only one variable we show - in one iteration - how EM works. Our first example consists of 10 data, from which two (cases 5 and 7, marked by xx) are missing:

80 – 70 – 60 – 55 – xx – 65 – 50 – xx – 40 – 70 – 45 – 30

The mean of the complete sample is 59.50, with $SD = 12.349$. The mean of the sample with the two missings is 61.250, $SD = 11.573$.

- First step: Estimation of the parameter(s) assuming a normal distribution: $\mu' = 61.250$, $\sigma' = 11.573$ (we remind the reader that $\bar{x}$ or μ' is a Maximum Likelihood, and a Least Squares, estimator of μ, see Section 9.2).
- Second step: Use the estimated parameter to fill in (= to impute) the missing data. Thus we fill in 61.250 (in case 5 and case 7) twice.
- Third step: Reestimate the parameters: $\mu' = 61.250$, $\sigma' = 10.206$.
- Fourth step: Use the reestimated parameter to fill in (= to impute) the missing data: We now fill in 61.250 (in case 5 and case 7) twice.

  Convergence reached? yes, as the imputed data do not change any more.

The EM approach for more complicated data structures is more difficult to explain, but the principle should be clear by now. We also see that EM (and the imputation by means, which is equivalent in this simple example) has one drawback: Standard errors tend to be too small, and as a consequence α levels are also too small.

We give another example, using the bivariate data set discussed earlier. The EM algorithm starts by imputing data based on a regression equation and added residuals. Thus, after 1 iteration the imputed value for case 5 is

62.371, which is close to the value found by the regression approach without adding a residual. When we set the maximal number of iterations at 25, in which the regression parameters are reestimated, we finally obtain a value of 62.247 for case 5 and 47.047 for case 7.

We now apply the EM algorithm to the data of Table 9.10b with repeated measures. This data set is more complicated, as three 'variables' are involved (the data obtained at three moments in time: t_1, t_2 and t_3). Multivariate distributions and multivariate regression analysis, in which covariance matrices (see Appendix B) play a role, are ingredients of the estimation process. That is why we refrain from giving the successive steps and intermediate results. After 1 iteration the imputed values are 3.99 for $x_{2.2}$ and 5.63 for $x_{4.3}$. After a maximum of 25 iterations the imputed values are 3.06 and 6.94 respectively; the original values were 2 and 7. The resulting $F_{2,6}$ ratio is 19.78, $p = 0.001$, still close to the values obtained for the complete data set. The EM algorithm we used is implemented in the MVA procedure: Missing Values Analysis of SPSS. Using the procedure one can choose the variables to which the MVA should be applied. For instance, if one has a design with two within-subject factors, of which one - let us call it 'words' - is nested within the other - for instance 'type of words' - (a default design in psycholinguistics), it is obvious that only the variables which refer to the words of one Type of words are used in the estimation process. Thus, the imputation process should be carried out in different stages, taking the structure of the data set into account.

The *seventh* imputation procedure is *multiple imputation*. It is gaining increasing popularity among statisticians and researchers, although at this moment in time (2004) it is not yet available in many statistical software packages. The approach can be seen as an attempt to adjust for the underestimation of standard errors, which is often the result of imputation. A simple bivariate example, used to estimate the correlation coefficent ρ, may suffice to explain the principle of multiple imputation. Assume we have a criterion variable y and a predictor x. There are missing data on y, but not on x. In the regression approach estimates of the regression coefficients are calculated and the resulting regression equation is used to fill in the missing values: $y_i' = B_0 + B_1 \times x_i$. The final step is to estimate the value of the estimator of ρ on the basis of all data, with the imputed values included. There are two problems with this approach:

- The standard deviation of the values of y is underestimated, as no random error is present in the imputed values of y. The formula for the correlation coefficient $r_{x.y}$ is $\text{cov}(x,y)/(\sqrt{s_x \times s_y})$. An underestimate of the s.d. of y means an overestimate of ρ.

- The estimator of the standard error of the parameter at issue (here the correlation coefficient) is too low, as the values of y do no vary enough.

The first problem can be remedied by adding random variation; this can be obtained from the distribution of the residuals of y. These residuals are simply the differences between the observed values of y and the predicted values of y, i.e. y'. Random draws of these differences can be added to the imputed values of y. The result will be a larger value of s_y; thus we will get a smaller value of $r_{x.y}$, which is closer to ρ.

The second problem, that of an estimator of the standard error which is too small, has not been solved yet. The multiple imputation approach aims at dealing with this problem. A number of imputations are carried out and at the same time random variation is added. Allison (2002) showed that a better estimator of the population parameter (here ρ) can be obtained if we take the mean of the estimators of ρ calculated in the different imputation runs. The standard error of the mean of the parameter *par* (here $\bar{r_{xy}}$) can be used with the following formula (cf. Rubin 1987):

$$SE(\overline{\text{par}}) = \sqrt{\frac{1}{M}\sum_k s_k^2 + (1+\frac{1}{M})(\frac{1}{M-1})\sum_k(\text{par}_k - \overline{\text{par}})^2} \qquad (14)$$

with M = number of imputation replications, σ_k^2 = the estimated standard error of $\bar{p}ar$. The use of this formula will lead to a better estimator of the standard error of the parameter at issue.

Our overview of procedures for repeated measures is not complete; there are more approaches available, like *raw maximum likelihood*, also called *direct maximum likelihood*. For more details we refer to Allison (2002), this also for the software needed, as many standard statistical packages only offer a restricted range of procedures for the analysis of missing data. With regard to practical recommendations for the analysis of missings in repeated measures designs we think that the procedure suggested by Allan & Wishart is a straightforward approach, as both the values observed for the variable and the values of the participant play a role in the imputation procedure (see formula (15)). A possible drawback is the necessity to carry out the imputation with ad-hoc software. The hot-deck procedure can be useful in hierarchical designs, in which the nested factor provides similar cases from which random samples can be drawn. Here too the drawback is the necessity of dedicated software. This does not hold for the EM-algorithm or the regression approach implemented in most software packages. Choosing the relevant variables for the estimation process is an important issue and the tendency of EM to underestimate standard errors.

> – Give four approaches to missing data in analysis of variance.
> – What is the difference between MCAR and MAR?

9.6. Suggestions for statistical reporting

EXAMPLE 1:
A one-way analysis of variance was carried out on the scores obtained in the three randomly selected schools. As the number of participants in each school varied we opted for the Maximum Likelihood approach provided by the procedure Mixed Models of SPSS (cf. Westfall (1987)). The ML estimator of σ^2 of the mean scores obtained in the schools was 4.437, and the value of the associated Wald statistic was not significant: $z = 0.167$, $p = 0.868$. We have no reason to assume that the scores of students obtained in the different schools differ.

EXAMPLE 2:
An analysis of variance was carried out on the Reaction Times obtained from 30 participants. Both the factors 'word type' (3 levels) and 'word' (5 levels, nested within 'word type') were considered fixed factors. 11.7% of the reaction times were missing as they were either shorter than 100 or longer than 1200 ms; consequently they were excluded from the analysis. In order to impute the missing data we opted for an Maximum Likelihood approach, in which the imputation process was repeatedly applied for the RTs obtained in each of the three word types. As the original degrees of freedom (levels of 'word type' minus 1 for the numerator of the F ratio and (levels of 'Type' minus 1) $\times$ Number of subjects minus for the denominator) are not affected by the imputation procedure, no adjustment of df was applied. The 'word type' factor was significant, $F_{2,58} = 12.34$, $p < 0.01$, $MSE = 5.450$.

9.7. Terms and concepts

- *Maximum Likelihood (ML) estimator*: An ML $\hat{\theta}$ estmate of a parameter θ is an estimator that maximizes the likelihood of getting the sample that is actually observed. For instance, if 7 out of 10 subjects have a 'uvular' /r/ (instead of an 'apical' /r/), the ML estimator of p to observe 'uvular' /r/ is $\hat{p} = 0.70$. This estimator yields the highest probability of getting the observed data. We can check this by assigning different values to $\hat{p}$. If this is 0.60, the probability of getting the observed percentage is, by using the binomial theorem: $(\frac{10}{7}) \times 0.60^7 \times 0.40^3 = 0.215$. By setting $\hat{p} = 0.70$ this probability increases to 0.267. With the value $\hat{p} = 0.80$ it decreases to 0.201. Thus, the maximum likelihood estimator in this sample is $\hat{p} = \frac{n_i}{N}$.

- *EM: expectation maximalization*: An imputation procedure which consists in iteratively estimating parameters of the distribution(s) from which the data are obtained, filling in the missing data on the basis of these distributions, re-estimating the parameters of the distribution(s) at issue on the basis of the data set including the imputed values, until the missing values do not change any more.
- *Imputation*: If missing data are filled in on the basis of a specific rule, for instance: use the mean of the observed values, one says that the missing data were imputed.
- *Multiple imputation*: A procedure which consists of obtaining a number of data sets which are the result of imputing and adding random variation. SES are based on the different data sets.

9.8. Exercises

1. Compare the ANOVA, ML, and REML approaches to the following one-way data with 'group' as fixed factor:

group		
I	II	III
1	2	3
7	1	6
8	2	9
4	3	8
6	1	7

2. Calculate the ML estimator of σ_α^2 of the 'group' factor of Exercise 1.
3. Carry out the ML procedure on the data of Exercise 1 for the situation in which the first two data of each group are assigned to the first level of another factor (factor 2), and the other two data to the second level of that factor. Compare the values of AIC for the one-way and two-way design.

4. The following data set is given, with four subjects and two within-subject factors: 'type of words' (two levels), and 'words' (nested within 'type of words').

	T_A			T_B		
	W_1	W_2	W_3	W_4	W_5	W_6
S_1	600	630	650	630	640	660
S_2	590	610	600	600	620	630
S_3	620	600	580	640	610	600
S_4	580	620	620	600	630	630

a) Carry out an analysis of variance, assuming 'words' is a fixed factor; use the syntax mode of SPSS.

b) Assume two observations are missing (S_1,T_A,W_1) and (S_3,T_B,W_3). Apply the MVA procedure of SPSS, options ME and Regression, to impute data to the missings, and carry out GLM, Repeated Measures to the two data sets thus obtained.

Chapter 10
Alternatives to analysis of variance

10.1. Preview

In this final chapter we discuss three quite recent alternatives of the techniques which were the subject of the preceding chapters: a) randomization tests, b) bootstrap procedures, and c) multilevel analysis.

Both *randomization tests* and *bootstrap procedures* provide nonparametric alternatives to ANOVA. In Section 10.2 the randomization test is introduced. This test is particularly suitable in situations in which specific assumptions of *t* tests and ANOVA are not met. In Section 10.3 we give a short overview of bootstrap procedures. Bootstrap procedures are, in a way, similar to randomization tests as they use self-generated distributions (that is why both procedures are called 'resampling' techniques). They are particularly useful when assumptions of standard procedures are not met. Both randomization tests and bootstrap approaches are not (yet) available in many statistical packages, or in a limited number of applications. If one wants to use SPSS for randomization tests, one has to resort to syntax, which is made available in a number of books, or on the web. For bootstrapping one has to use dedicated software, which can often be downloaded from the web. Although randomization tests and bootstrap procedures are not frequently used in the current literature on language research, we expect that researchers will find references to these approaches in the not too distant future, as they will be implemented in most statistical software packages.

Multilevel analysis has become popular especially in applications with repeated measures. It is able to deal with heterogeneous growth curves in longitudinal research in a way ordinary analysis of variance is not. In Section 10.4 we present an example of a multilevel approach to a longitudinal design, and compare the results with an ordinary analysis of variance repeated measures design.

10.2 Randomization tests

The main focus of this book is on parametric statistics. This means that the statistical tests and techniques we discuss are based on well-known theoretical distributions and particularly the normal, F and t distributions, which implies that the dependent variable is measured at the interval level.

An interesting alternative is provided by randomization tests which do not resort to theoretical distributions of statistics like F and t. Instead, the distribution of a statistic is derived from the observed data themselves. A randomization test assesses how often the statistic in question takes a value in the self-created distribution equal to or greater than the magnitude of the statistic in the original sample. The distribution of the statistic is generated by permuting the observed values over the treatments and calculating the value of the statistic each time. The advantage is that no assumptions on the form of the distribution in the population are necessary. For this reason these techniques are well suited to situations where assumptions like normality and homogeneity of variances are evidently violated. Prerequisite to the procedures is the availability of calculation power: The distribution of the statistic has to be derived from the original sample by carrying out a large number of permutations and (re-)calculations of the test statistic. The wide-spread availability of powerful computers has promoted the use of the procedures under discussion, and it is now generally believed that randomization tests will soon be standard tools of the researcher's statistical equipment.

The principles of a randomization test can be illustrated on the basis of the small data set given in Table 10.1. It represents a one-way design involving two treatments.

Table 10.1. Data for two treatments, A_1 and A_2

	A_1	A_2
	1	3
	2	9
$\bar{X}_j$	1.5	6.0

If a t test for independent samples - equivalent to a one-way ANOVA - is applied to the data of Table 10.1, the t value is 1.80 with $df = 1$ and a p value of 0.323. This outcome is clearly not significant at the 5% level. The assumption of homogeneity of variances, on which the theoretical t distribution is based, is not met: $s^2_{A_1} = 0.499$ and $s^2_{A_2} = 18.003$. If we go on using this theoretical distribution and make precise statements on the probabilities of occurrence or non-occurrence of our findings in the case of a true H_0 hypothesis, we wrongly suggest precision and accuracy in our conclusions.

As said before, randomization tests create their own distributions. To this end the observations are assigned to the treatments in all possible ways, by means of a so-called *permutation procedure*. For a two-treatment design there are $N!/(n_1! \times n_2!)$ possible permutations. In our example the number of possible permutations is $4!/(2!\times 2!) = 6$ as shown in Table 10.2. It is important to note that in a technical sense these permutations are combinations, as the order of the elements within a sample is not relevant. That is why some authors (cf. Todman and Dugard 2001) prefer to use the word *arrangements*.

Table 10.2. All possible permutations of the original four scores of Table 10.1, with associated means per 'treatment'

	permutations					
	1	2	3	4	5	6
	1 3	1 2	1 2	3 1	2 1	2 1
	2 9	3 9	9 3	9 2	9 3	3 9
$\bar{X}$	1.5 6.0	2.0 5.5	5.0 2.5	6.0 1.5	5.5 2.0	2.5 5.0

We can calculate a t value for all arrangements, but the simple difference $|\bar{X}_{A_1} - \bar{X}_{A_2}| = |D|$ as a statistic will do. The distribution of the magnitudes of the test statistic, be it t or D, is then used to determine the significance of the observed value of the test statistic. The two-tailed probability associated with the difference observed in our small-scale experiment is the proportion of data permutations resulting in a value of $|D|$ or $|t|$ greater than or equal to the difference observed in the experiment. The observed value of $|D|$ is 4.5; a value of $|D| \geq 4.5$ occurred in 2 of the 6 permutations, so p (two-tailed) is $2/6 = 0.33$. The magnitudes of $|t|$ are 1.80, 1.40, 0.71, 1.80, 1.40, 0.71. Thus a $|t| \geq 1.80$ occurred in 33.3% of the cases in the permutation set.

With this procedure we test the null hypothesis that the scores of every subject (experimental unit) are independent of the treatment. If H_0 is correct, a random assignment of the scores will lead to two values of $|D|$ as large as or larger than the value of 4.5, i.e. in 33.3% of the cases. Thus, we cannot provide strong evidence against the H_0; obviously our result is not significant at the 5% level.

Possibly to the suprise of the reader, we used the simple absolute difference between the means $|D|$ as a statistic in our two-treatment example. The reason is that the rank orders for $|t|$ and $|D|$ are the same, and consequently the same p is found for $|D|$ and $|t|$; $|D|$ and $|t|$ are said to be equivalent test statistics. The equivalent statistic for F is $\sum T_j^2/n_j$, where T_j is the total of the scores in treatment j, and n_j is the sample size.

The same procedure as sketched above for a mini-example with two treatments can be applied to larger data sets and to more complex designs. At first sight there is a drawback: The number of arrangements which have to be computed increases rapidly. For a one-way analysis of variance the number of arrangements is:

$$\frac{N!}{n_1! \times n_2! \times \ldots \times n_k!} \qquad (1)$$

with N = total number of observations, and n_j = number of observations in treatment j. For a one-way design with three treatments and 6 observations per treatment the number of arrangements is $18!/(6! \times 6! \times 6!) = 331{,}776$. Fortunately, we are not obliged to compute all of them the way we did in the introductory example. Instead of forming the complete set of arrangements one can resort to the *random permutation method.*

In the random permutation method all data are assigned an index; a random number generator selects n_1 numbers (without replacement) from N index numbers and assigns the corresponding scores to treatment 1; the same procedure is carried out for the other treatments drawing n_j numbers. The whole procedure is repeated a great number of times (1000 will suffice in many situations), and each time the statistic is calculated and stored. The (1000) values of the statistic constitute a distribution from which the p value is derived. This p value is obtained by determining the frequency with which the values of the statistic are equal to or exceed the observed value. The p value is calculated by dividing this number by the total number of permutations.

The randomization test can be exemplified by a concrete application, based on the data given in Table 10.3, reproduced from Table 7.1. In Chapter 7 we showed that the assumption of homogeneity of variances of these data was not warranted (Levene's test was significant at the .032 level). This violation renders the analysis of variance questionable in light of the small samples. The data are given in Table 10.3.

The results of an analysis of variance of the data of Table 10.3 are presented in Table 10.4. The treatment effect is almost significant ($p = .0525$).

For our dataset the F ratio is 39.467/10.233 = 3.857. The ratio is made up of two Mean Squares. The numerator is MS_{between}, the denominator is MS_{error}, which in turn is the result of SS_{error} divided by the degrees of freedom.

Table 10.3. Data obtained in three groups with five participants each; reproduction of Table 7.1

	G_1	G_2	G_3
	4	2	3
	5	3	7
	7	4	6
	10	3	2
	16	2	8
$\bar{X}$	8.40	2.80	5.20
s^2	23.30	0.70	6.70

Table 10.4. (Part of the) SPSS output from an analysis of variance (Compare Means, One-Way ANOVA) applied to the data from Table 10.3

Source	Sum of Squares	df	Mean Square	F	Sig.
Between Groups	78.933	2	39.467	3.857	.051
Within Groups	122.800	12	10.233		

In Chapter 7 we demonstrated the effect of a log transformation of the original variables in an attempt to obtain homogeneous variances. Alternatives are Welch's test (see Chapter 7, Section 3) but also a randomization test or bootstrapping. The rationale of randomization tests is to calculate a statistic for each arrangement of data and to find out how often the obtained statistic is equal to or larger than the statistic obtained for the actual dataset. In our case the F ratio is the conventional statistic; it involves the calculation of two values: MS_{between} and MS_{error}. An equivalent statistic is SS_{error}. If SS_{error} decreases, MS_{error} decreases as well, and MS_{between} increases. Thus, by calculating the number of times that an SS_{error} is found at least as small as 122.800 in 2000 or 5000 randomly chosen arrangements, we can assess the significance of the result obtained for our actual data set. Randomization tests are not yet included in most standard statistical packages. Todman and Dugard (2001) provide syntax for three packages: MINITAB, EXCEL and SPSS which enables researchers to carry out randomization tests on data obtained in a number of designs. By default 2000 different arrangements are used for each design. With very small data sets, like the first dataset presented in this section, for which only six different arrangements are possible, the number of 2000 constitutes a kind of 'overkill'. But for more realistic datasets, 2000 is a sufficient sample of all possible arrangements. The test statistic used in randomization tests is not always the one we are used to, but just an equivalent version of the more traditional test statistic. In Table 10.5 we give the output of the SPSS syntax provided by Todman & Dugard; 5000 arrangements were used.

Table 10.5. Output of the SPSS syntax provided by Todman & Dugard for randomization tests; 5000 arrangements were used for a randomization test on the data of Table 10.3

```
RSS
   122.800

count of arrangements RSS at least as small
   48

probability
   10 ** -2 x 2.449
```

RSS in Table 10.5 means 'Residual Sum of Squares' (= SS_{error}); the *RSS* of the observed data set is 122.800. In the 5000 arrangements of the data 48 ones resulted in an *RSS* of 122.800 or less: 2.45% (equivalent of $10^{-2} \times 2.449 = (1/10^2) \times 2.449$). Consequently we have reason to reject the null hypothesis.

The differences between randomization tests (RT) and parametric tests (PT) can be summarized as follows:

1. In randomization tests there are no requirements concerning the parameters of the populations from which the samples are drawn, like normality, homogeneity of variances etc.
2. In RT there is no need to have randomly drawn samples from one or more populations, although random sampling facilitates the interpretation of the results.
3. In RT there are no precise null hypotheses framed in population parameters like $\mu_1 = \mu_2$. The hypotheses being tested can be formulated as: the scores obtained by participants are independent of the treatment they were assigned to.
4. In both RT and PT random assignment of participants to treatments/conditions is crucial.
5. Randomization tests are not concerned with estimating population parameters; thus point estimations with Confidence Intervals are beyond

the realm of randomization tests.

6. The form of the test statistic is quite often less complicated than in parametric tests where, for instance, differences between treatment means are 'scaled' by estimates of standard errors of the differences. In randomization tests a difference between two treatment means suffice.

10.3. Bootstrapping

The technique of bootstrapping was specifically designed to estimate the stability of estimated parameters (Efron and Gong 1983). Bootstrapping can be used to estimate the sampling variability of the estimates of parameters. The underlying idea is simple. Assume that we would like to estimate the mean of a random variable by calculating the mean of a sample of size n. The standard error of the mean is equal to the standard deviation σ of the distribution in the population divided by n; σ is estimated in its turn by s (see Chapter 2). The standard error of the mean is then estimated by:

$$SE = \sqrt{\frac{\Sigma(X_i - \bar{X})^2}{n-1}} / \sqrt{n} = \frac{s}{\sqrt{n}} \qquad (2)$$

This well-known method of calculating the accuracy of an estimator like $\bar{X}$ cannot be applied to other estimators like the median when the population is not normally distributed. In those cases the bootstrapping method can help. This technique boils down to a simulation of the distribution of the estimator by randomly drawing – with replacement – a great number of samples of size n from the original sample of n observations. These samples are called *bootstrap samples*. For each sample which is obtained in this way the required parameter, for instance the mean or the median, is calculated. Given this set of sampled values of the parameter in question the standard error can be calculated. The standard error (SE) of the parameter can be approximated as follows:

$$SE_{\text{bootstrap}} = \sqrt{\frac{\Sigma_{i=1}^{b}(\theta_i^* - \hat{\theta})^2}{b-1}} \qquad (3)$$

where θ_i^* = the estimated value of the parameter in bootstrap sample i, $\hat{\theta}$ = the value of the parameter in the original sample, and b = number of bootstrap samples = number of random samples with replacement.

By drawing a great number of bootstrap samples the standard deviation of the resulting distribution provides a good estimate of the standard error of the relevant parameter. At first it may seem rather strange that by resampling the original sample an analogon of the population distribution can be created. A prerequisite to this analogy is that we know how the original sample was collected. For instance, when bootstrapping is applied to establish a confidence interval for a correlation coefficient, we must not forget that the data in correlation problems are pairs of observations. The *sampling scheme* of the original sample defines the sampling scheme in bootstrapping. The original sample – taken from an unknown population – serves as a known population from which the bootstrap samples are drawn. The probability distribution of the samples which can be drawn is not altered by the bootstrap draws as the requirement of sampling with replacement, an important characteristic of bootstrapping, ensures that the probabilities of the samples remain equal. This is exactly what happens when we draw samples from a large unknown population.

In order to illustrate the basic principles of bootstrapping we are going to present a simple example, based on a small sample drawn from a normally distributed population with $\mu = 10$, and $\sigma = 4$. Two aspects of bootstrap samples are demonstrated:

1. The standard deviation of the bootstrap distribution – the latter is symbolized by F^* – can serve as an estimator of the *SE* of $\bar{X}$, the mean of the original sample. N.B. F^* has nothing to do with the F ratio.
2. A confidence interval (= *CI*) can be calculated on the basis of F^*.

In the subsequent example the *CI* is used to illustrate the application of bootstrapping to test hypotheses. In order to be in line with formula (3) given above and the literature on this topic we will use the following symbols:

$\hat{m}$	=	the estimate of a population mean based on the mean of the original sample;
m_i^*	=	the mean of a bootstrap sample;
F^*	=	a bootstrap distribution.

In the first example we focus on the *SE* of the arithmetic mean of a sample. Table 10.6 contains the original sample with 15 observations.

Just three hundred bootstrap samples were drawn from this original sample. For clarity's sake we reproduce three of these samples in Table 10.7,

Table 10.6. A sample of 15 cases from $N(10,4)$; $\hat{m} = 10.333$, $s = 3.933$, $SE = 1.013$ (estimated by $s/\sqrt{n}$)

11, 15, 10, 11, 5, 8, 13, 11, 7, 18, 11, 6, 4, 15, 10

obtained by drawing with replacement. On the basis of the means of the bootstrap samples – symbolized by m_i^* – a bootstrap distribution is obtained, which is reproduced in Figure 10.1.

Table 10.7. Three bootstrap samples with associated m_i^*s

sample 1	sample 2	sample 3
15	4	15
11	5	18
11	10	15
10	11	11
11	6	15
5	6	15
7	15	18
11	8	5
11	13	11
4	10	6
13	4	7
11	6	11
4	10	6
6	15	10
10	18	15
$m_1^* = 9.33$	$m_2^* = 9.40$	$m_3^* = 11.87$

The mean of this distribution, viz. m^* is 10.292; it approximates the mean of the original sample, which was 10.333. Bootstrap theory states that the standard deviation of this distribution is an estimate of the *SE* of $\bar{X}$, the mean of the original sample and symbolized by $\hat{m}$, the estimate of the population mean. In our case the $s = .927$, whereas the analytically based estimate of the *SE* is 1.013 (see Table 10.6).

The confidence interval is an important concept in bootstrapping. There are different methods available to calculate a *CI*. One of these is the percentile method, in which those percentiles of the bootstrap histogram are applied which correspond to the $(1 - 2\alpha)$ confidence interval. If the 90% *CI*

COUNT	MIDPOINT	ONE SYMBOL EQUALS APPROXIMATELY 1.25 OCCURRENCES
0	7.3	
2	7.6	***
1	7.9	*
0	8.2	
6	8.5	********
8	8.8	**********
22	9.1	*****************************
33	9.4	***
28	9.7	************************************
40	10.0	**
30	10.3	***************************************
30	10.6	***************************************
33	10.9	***
26	11.2	**********************************
15	11.5	*******************
17	11.8	**********************
5	12.1	******
2	12.4	***
1	12.7	*
1	13.0	*
0	13.3	

Figure 10.1. Histogram of the values of parameter m_i^* for the data of Table 10.6

has to be calculated for instance, the 5th and 95th percentiles of the bootstrap distribution are used. In our case this yields the interval 8.933–11.800.

A simple example of hypothesis testing based on bootstrapping can be given for the data in Table 10.3, which we already used to demonstrate randomization tests and which is a reproduction of Table 7.1. It is a one-way design with three treatments (fixed effects). The problem with these data was the heterogeneity of variance in the treatments. We applied bootstrapping on these data with the free RESAMPLING program developed by Howell (www.uvm.edu/ dhowell/StatPages/Resampling/ResamplingPackage.zip). For bootstrapping the program uses F ratios. The crucial difference with a parametric one-way test in which F ratios are used is, that in the bootstrap approach the F ratios are not compared with values of the theoretical probability density function (distribution) of F, but are used as indices of the magnitude of the ratio of the between-group and within-group variations. If the bootstrapping yields only a small proportion of F ratios equal to or larger than the actual observed F ratio, one may conclude that there is reason to reject the H_0, which says that both $MS_{between}$ and MS_{within} estimate the same variance, i.e. σ^2_{error}. We ran 5000 bootstraps and obtained a percentage of 4.2% of F values equal to or greater than the observed F ratio of 3.857. The result is significant at the 5% level. In Figure 10.3 the distribution of F ratios obtained in the 5000 bootstraps is depicted.

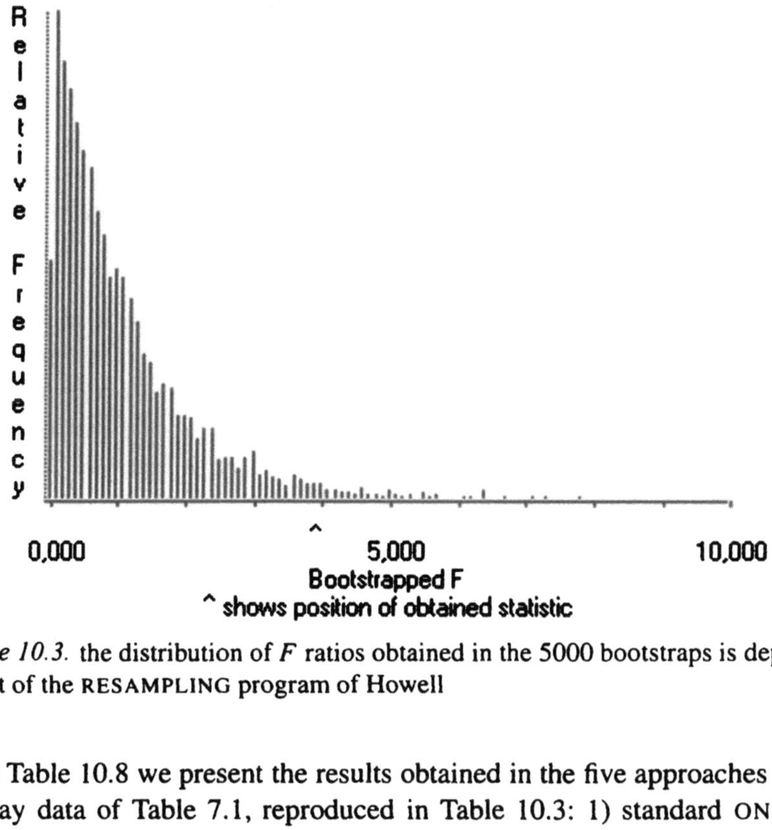

Figure 10.3. the distribution of F ratios obtained in the 5000 bootstraps is depicted; output of the RESAMPLING program of Howell

In Table 10.8 we present the results obtained in the five approaches to the oneway data of Table 7.1, reproduced in Table 10.3: 1) standard ONEWAY ANOVA, 2) oneway anova carried out on the logtransformed data, 3) Welch's test, 4) randomization test, and 5) bootstrap analysis.

Table 10.8. Levene's test plus results obtained in five approaches to the data from Table 10.3

	F	df_1	df_2	sig.
Levene's test	4.666	2	12	.032
standard one-way	3.857	2	12	.051
one-way on log transformed data	5.086	2	12	.025
Welch	4.663	2	5.972	.065
randomization	–	–	–	.024
bootstrap	–	–	–	.042

Both the original oneway analysis and Welch's test were not significant at the 5% level, whereas the analysis carried out on the logtransformed data and both the randomization and bootstrap approach yielded significant results.

10.4. Multilevel analysis

Quite recently a technique has become available which is frequently used as an alternative both to multiple regression analysis and analysis of variance: Multilevel analysis (ML), also known as Hierarchical Linear Modelling (HLM). Especially in the analysis of repeated measures this alternative is more powerful and more flexible.

In the preceding chapters we presented analysis of variance and multiple regression as techniques which can be used to detect effects of within-subject factors. In many research settings the within-subject factors reflect individual change which may or may not be affected by between-subjects factors, such as 'gender' or 'socio-economical status'. Often the levels of the within-subject factor(s) represent different points in time, and the model equations underlying individual change in time are, in fact, growth models. The model equation is assumed to incorporate a number of relevant factors: the participants' personal bias (a random factor manifesting the unobserved heterogeneity), the between-subject factor(s), the within-subject factor(s) and an interaction of between and within-subject factors. If the latter turns out to be significant, we have reason to assume that the between-subject factor(s) affected the time course of the scores (the within-subject factor(s)). However, this approach has a number of limitations, associated with three assumptions underlying ANOVA with repeated measures:

- Equal variances: The sphericity assumption states that the variances of the differences between the levels of the within-subject factor(s) are equal. This assumption is not realistic, as we may expect that participants will diverge in their scores in the course of time (cf. Huynh and Feldt 1976).
- Equal growth rates: A related phenomenon is that participants tend to exhibit individual growth rates, which may be a function of categorical or continuous background (indicator) variables. In ANOVA these individual growth rates cannot be incorporated in the model equation.
- Equal time spacing: Spacing between measurement occasions and the number of occasions are assumed to be equal. Especially the last as-

sumption (no missing data) is awkward, as in many research settings missing data do occur (see Chapter 9).

Multilevel analysis allows researchers to study variation in growth rates as a function of background variables such as treatment and socio-economical status. We present one option of this approach, which is called the intercept-and-slopes-as-outcomes model; this option assumes a linear model for growth rate. In this approach two levels (or stages) are distinguished, each characterized by its own model:

- Level-1: A *within-subject* model which models the individual growth trajectory.
- Level-2: A *between-subject* model, in which the parameters of the level-1 model are modelled as a function of background characteristics.

We will start the discussion of this technique on the basis of a simple linear regression equation. There is one outcome variable, Y_i, for instance the scores on a reading test, and one predictor variable, the socio-economical status of the parents (= SES): X_i.

$$Y_i = \beta_0 + \beta_i X_i + \varepsilon_i \qquad (4)$$

where:

- β_0 = the intercept. It is the expected achievement of a student with parents whose SES is zero.
- β_1 = the slope. It is the expected change in achievement associated with a unit change in SES.
- ε_i = the error term. It in fact represents a unique effect associated with student i; ε_i is normally distributed: $N(0, \sigma^2)$.

Quite often it is a good idea to center the independent variable by subtracting the mean from each score: $X_i - \bar{X}$, because plotting the resulting regression line yields a β_0 which corresponds with – in our case – the mean achievement score.

Let us assume that there is not just one school, but two. We have two regression lines, obtained for students at these two schools. Two observations can be made:

- $\beta_{01} > \beta_{02}$: Students in school 1 are higher achievers than students in school 2.
- $\beta_{12} > \beta_{11}$: SES has less influence on the achievement scores in school 1

than in school 2.

The example given above suggests that each school has its own slope and intercept. Thus equation (4) becomes:

$$Y_{ii} = \beta_{0j} + \beta_{ij}(X_{ij} - \bar{X}_{.j}) + \varepsilon_{ij} \qquad (5)$$

The subscript j is added as the regression equations for students in schools j can differ. It may even be the case that the values of intercepts and slopes are determined by a background (indicator) variable W, for instance the size of the schools, small vs. large: factor 'Size'. As a consequence, the intercepts and slopes can be modelled on the basis of that indicator variable:

$$\beta_{0j} = \gamma_{00} + \gamma_{01}W_j + u_{0j} \qquad (6)$$

and

$$\beta_{1j} = \gamma_{10} + \gamma_{11}W_j + u_{1j} \qquad (7)$$

In equations (6) and (7) two linear regression equations are given, in which γs are introduced instead of the conventional β's, in order to avoid confusion with the βs which have to be modelled:

- γ_{00} = mean achievement of students with parents who have an average SES
- γ_{10} = average effect of SES on the achievement of the students
- γ_{01} = average effect of 'size' on mean achievement of students with parents with an average SES
- γ_{11} = the average change of the slope as a function of the indicator variable 'size' of schools
- u_{0j} = term which reflects random variation of β_{0j} around γ_{00}
- u_{1j} = term which reflects random variation of β_{1j} around γ_{10}.

Equation (5): $Y_{ij} = \beta_{0j} + \beta_{1j}(X_{ij} - \bar{X}_{.j}) + \varepsilon_{ij}$ is referred to as a level-1 equation. It is the equation which models the behaviour of individuals, in our case of student i in school j.

Equations (6) and (7) - $\beta_{0j} = \gamma_{00} + \gamma_{01}W_j + u_{0j}$ and $\beta_{1j} = \gamma_{10} + \gamma_{11}W_j + u_{1j}$ - are referred to as level-2 equations. The latter model the intercepts and slopes of the students in j different schools on the basis of the indicator variable W. Thus two different levels are distinguished, each with its own regression equations. This is where the term multilevel analysis got its name from. The equations given in (5), (6) and (7) can be combined in one single equation as follows:

$$\begin{aligned} Y_{ij} &= \gamma_{00} + \gamma_{01}W_j + \gamma_{10}(X_{ij} - \bar{X}_{.j}) + \gamma_{11}W_j(X_{ij} - \bar{X}_{.j}) + \\ &\quad u_{0j} + u_{1j}(X_{ij} - \bar{X}_{.j}) + \varepsilon_{ij} \end{aligned} \tag{8}$$

Rearranging terms we can distinguish between two parts, the fixed part and the random part; the latter only contains terms which refer to components which are assumed to vary randomly:

- Fixed part:
  $\gamma_{00} + \gamma_{01}W_j + \gamma_{10}(X_{ij} - \bar{X}_{.j}) + \gamma_{11}W_j(X_{ij} - \bar{X}_{.j}) + u_{0j} + u_{1j}(X_{ij} - \bar{X}_{.j})$
- Random part:
  $u_{0j} + u_{1j}(X_{ij} - \bar{X}_{.j}) + \varepsilon_{ij}$

We illustrate the multilevel approach on the basis of a simple data set, consisting of 'fluency' scores obtained from stutterers with two levels on the background variable 'education' (= edu). Stuttering was recorded at 4 moments in time (t_i). The data are given in Table 10.9.

Table 10.9. Fluency scores obtained from stutterers distinguished in two educational groups ('edu')

edu	t_1	t_2	t_3	t_4
	4	5	6	2
	3	6	5	3
0	2	4	3	2
	4	5	6	4
	3	4	5	5
	4	7	8	7
	3	5	7	6
1	4	7	9	8
	4	8	7	7
	3	5	7	6

The details of the dataset in Table 10.9 are as follows:

- 10 participants, measured at 4 moments in time
- within-subjects variable: 'time' (4 levels)
- between-subjects variable: 'edu' (2 levels, coded as 0/1)
- no centering of the dependent variable.

Level-1 model:

$$\mathrm{STUT}_{ti} = \beta_{0i} + \beta_{1i}(\mathrm{time})_{ti} + \varepsilon_{ti} \qquad (9)$$

with:

- STUT_{ti} = 'stutter-score' of participant i at time t
- β_{0i} = true score of participant i at time 0 (intercept)
- β_{1i} = growth rate of participant i over the observation interval
- $(\mathrm{time})_{ti}$ = time of measurement of person i at time t and spacing of measurements may vary across persons)
- ε_{ti} (in the output of the software package HLM: R) = random error associated with participant i at time t.

Level-2 models:

$$\begin{aligned} \beta_{0j} &= \gamma_{00} + \gamma_{10}(\mathrm{edu})_i + u_{0i} \qquad (10) \\ \beta_{1j} &= \gamma_{01} + \gamma_{11}(\mathrm{edu})_i + u_{1i} \qquad (11) \end{aligned}$$

with:

- edu = background variable, indicated by W in formulas (6) and (7);
- γ_{00} = intercept;
- γ_{10} = average effect of 'Education' on the dependent variable;
- γ_{11} = average increase of the slope β_{1j} accounted for by differences in 'Education'
- u_{1i} = random error.

The growth rate parameters β_{0i} and β_{1i} can be estimated in different ways; we will not go into the details here. The two different estimates of growth rate are obtained at levels-1 and -2 respectively. At level-1 the βs are Least Squares estimates (OLS) based on separate regressions for the participants involved, whereas at level-2 the βs are estimated on the basis of the background variables W (method: Weighted Least Squares, not discussed here).

The significance of the gamma weights is tested with the t statistic; the degrees of freedom associated with t depend on the type of predictor variable. If it is a level-1 predictor, $df = N - f$, where f is the number of fixed effects in the model and N is the total level-1 sample size. For the level-2 predictors $df = J - p$, where p is the number of parameters in the equation and J is the level-2 sample size.

We carried out a multilevel analysis with the HLM 5 software package. In Tables 10.10a and 10.10b we present part of the default HLM output. The core of the output is the table with the estimation of the gamma coefficients and the associated t ratios ('T-ratio') and p values. There we can read which background variables (in our example there is only one background variable: 'EDU') affect the intercept of the growth rates ('BASE') and the SLOPE. It is important to bear in mind that there are three intercepts involved: β_0, γ_{00} and γ_{10}. HLM repeats the two models, level-1 and level-2, on the basis of the chosen input options, as given in Table 10.10a.

Table 10.10a. Part of HLM output: analysis of repeated measures data given in Table 10.8

```
The outcome variable is SCORE

  The model specified for the fixed effects was:
 ----------------------------------------------------

   Level-1                  Level-2
   Coefficients             Predictors
 ----------------        ---------------
   INTRCPT1, B0             INTRCPT2, G00
   TIME slope, B1           INTRCPT2, G10
                            EDU, G11

 Summary of the model specified (in equation format)
 ----------------------------------------------------

Level-1 Model
       Y = B0 + B1*(TIME) + R

Level-2 Model
        B0 = G00 + U0
        B1 = G10 + G11*(EDU) + U1
```

In the summary given by HLM the Greek letters β and γ are replaced by the Roman letters B and G respectively. After the summary the output of HLM proceeds and gives the Ordinary Least Squares (OLS) estimates of the intercepts and slopes of the level-1 model for the 10 different participants (Table 10.10b) and also gives estimates of the values of the coefficients of the fixed effects with the associated *SE*s, t and p values.

In Table 10.10b we see the values of the coefficients of the fixed effects, and the associated *SEs*, t and p values. There are two subtables related to the fixed effects; the difference resides in the estimation procedure. We only reproduce the first subtable.

Table 10.10b. Part of HLM output: analysis of repeated measures data given in Table 10.8

```
Estimation of fixed effects
 Level-1 OLS regressions
 ----------------------

 Level-2 Unit      INTRCPT1          TIME slope
 -----------------------------------------------------------
            1      5.50000          -0.50000
            2      3.50000           0.50000
            3      3.00000          -0.10000
            4      2.50000           0.70000
            5      2.50000           0.70000
            6      3.50000           0.80000
            7      4.00000           0.20000
            8      3.00000           0.60000
            9      3.50000           0.30000
           10      4.50000          -0.20000

The average OLS level-1 coefficient for INTRCPT1 =    3.32500
The average OLS level-1 coefficient for     TIME =    0.73000

 Final estimation of fixed effects:
 -----------------------------------------------------------
                                 Standard                Approx.
   Fixed Effect    Coefficient   Error     T-ratio  d.f.  P-value
 -----------------------------------------------------------
 For INTRCPT1, B0
    INTRCPT2, G00    3.325000  0.253766  13.103    19     0.000
 For TIME slope, B1
    INTRCPT2, G10    0.381219  0.128007   2.978    18     0.009
         EDU, G11    0.697563  0.126449   5.517    18     0.000
 -----------------------------------------------------------
```

The main question to be answered in this investigation was: is the background of the participants – 'education' – a relevant factor for the prediction of the scores obtained in the course of time. In analysis of variance (repeated measures) this question is answered by the absence/presence of a significant interaction between the within-subject factor 'time' and the between-subject factor 'education'. In our example this interaction is significant: $F_{3,24} = 8.466$, $p = 0.000$, (Analysis of variance with the GLM procedure of SPSS, repeated measures).

The multilevel approach yields the same result as can be seen in Table 10.10b. We have to inspect the coefficient and associated *t* test of EDU, under the heading 'TIME slope, B1'. This slope is the regression coefficient of β_1(TIME) which is modelled as $\beta_1 = \gamma_{10} + \gamma_{11}(\text{EDU}) + u_1$. The coefficient γ_{11} – called 'EDU, G11' in the output table – represents the weight which has to be attributed to the 'education' factor in order to predict the time course of the scores. The associated t_{18} value is 5.517, $p = 0.000$, a result which is equivalent to the result obtained in the repeated measures procedure of SPSS: The education level affects the time course of stutter scores.

The final subtable- - reproduced in Table 10.10c – gives the statistics associated with the three random effects: u_0 (U0) and u_1 (U1) which represent the random errors of the regression equations used to model the intercept and the growth rate respectively at level-2, and ε_{ti} (R) which is the random error at level-1.

Table 10.10c. Part of HLM output: analysis of repeated measures data given in Table 10.8

Final estimation of variance components:

Random Effect		Standard Deviation	Variance Component	df	Chi-square	P-value
INTRCPT1,	U0	0.19271	0.03714	19	14.90080	>.500
TIME slope,	U1	0.28459	0.08099	18	24.20064	0.148
level-1,	R	0.91315	0.83385			

Why should one use multilevel analysis when the results tend to be similar to the results obtained with analysis of variance? An important advantage of multilevel analysis is the fact that missing values do not result in missing cases. If an observation is missing in an analysis of variance (repeated measures), the whole case is deleted (listwise deletion), both in the MIXED approach of SPSS and in the GLM repeated measures analysis cf. Chapter 9). This is not the case with multilevel analysis. Thus, in situations in which missing data occur frequently (experiments with RTs, research with patients), Multilevel analysis is a good alternative to more conventional approaches (cf. Quene and Van den Bergh 2004).

10.5. Suggestions for statistical reporting

EXAMPLE 1:
In a randomization test of the hypothesis that the three contexts involved lead to different relative frequencies of assimilation, the proportion of 3000 randomly sampled data divisions yielding a test statistic (RSS = Residual Sum of Squares) at least as small as the experimentally obtained statistic was 0.032. Therefore, the obtained differences in the three contexts were statistically significant ($p < 0.05$).

EXAMPLE 2:
The speech quality scores obtained at four moments in time ('time' factor; fixed) by the patients in the two groups distinguished by their education level ('education'; fixed) were submitted to a multilevel analysis. There were two reasons to do so: a) The intervals between the four moments were different and b) The speech of three of the ten patients could not be recorded at one of the recording sessions. Two models were used, Level-1 and level-2. At level-1 the model is Score = B0 + B1*TIME+R, at level-2 it is B0 = G00 + U0 and B1 = G10 + G11*Group + U1. The 'group' factor significantly affects the time course of the speech quality scores. The weight G11 = 0.65, and the associated $t_{15} = 4.731$, $p < 0.01$. For group 1 the average time slope = 0.27 and for group 2 it is 0.33. Thus, the rate of improvement over time achieved by patients with a lower education level was lower than that achieved by patients with a higherlower level of education.

10.6. Terms and concepts

- *Randomization test*: A nonparametric test in which the sampling distribution is created on the basis of the original observations. To this end the observations are assigned to the treatments in all possible ways by means of a permutation procedure.
- *Bootstrapping*: A technique used to obtain a sampling distribution of an estimator by randomly drawing with replacement a great number of samples of size N from the original sample of N observations. Each time a bootstrap sample is obtained, the estimator is calculated. The large number of estimators constitutes a sampling distribution, which can be used to assess with which probability the observed value of the estimator – for instance $\hat{\mu}$ or the difference between two means – or a larger value could have been

observed. That probability is the equivalent of the ordinary α level.

- *Multilevel analysis*: An approach to the analysis of scores in which two or more levels are distinguished. Each level is modelled on the basis of regression equations. An example on the basis of longitudinal designs: participants who differ on a background variable (BACK) are measured at different moments in time (TIME). At level-1 the scores of a subject can be modelled as follows: $Y_{ij} = \beta_{0j} + \beta_{1j} \times \text{TIME} + \varepsilon$, at level-2 the regression coefficients of the equation at level-1 are modelled in turn: $\beta_{0j} = \gamma_{00} + \gamma_{01} \times \text{BACK} + u_{0j}$ and $\beta_{1j} = \gamma_{10} + \gamma_{11} \times \text{BACK} + u_{1j}$. The γ weights indicate to which extent the background variable BACK influences the time course of the scores obtained at the different moments in time.

10.7. Exercises

1. Carry out a randomization test on the following data:

C1	C2
1	10
2	15

2. In which of the two procedures – bootstrap and randomization – plays replacement a key role?
3. Does it make a difference whether the statistic in a randomization test for a oneway-design is an F ratio or just SS_{error}. Why?
4. How many arrangements are possible in a randomization test for a two-treatment design with five observations in each treatment?
5. Is it possible to determine how many bootstrap samples are possible for a two-treatment design with five observations in each treatment?
6. A multilevel analysis was carried out on data obtained in a design in which reaction times were observed for four types (TYPE) of words; each type contained 10 words. Observations were pooled over words. There were two groups of listeners, listeners with Dutch and listeners with British English as their mother tongue (MT). The following three regression equations are used:
   a) $\beta_{0j} = \gamma_{00} + \gamma_{01} \times \text{MT} + u_{0j}$,
   b) $Y_{ij} = \beta_{0j} + \beta_{1j} \times \text{TYPE} + \varepsilon$
   c) $\beta_{1j} = \gamma_{10} + \gamma_{11} \times \text{MT} + u_{1j}$
   Indicate which regression equations are called level-1 and which level-2?

References

Allan, F.E. and J. Wishart
1930 A method of estimating yield of a missing plot in field experimental work. *Journal of Agricultural Science* 20: 399–406.

Allison, Paul D.
2002 *Missing Data* Thousand Oaks, CA: Sage.

Anderson, Norman H.
1961 Scales and statistics. *Psychological Bulletin* 58, 305–316.

Anderson, D.R., D.J. Sweeney, and T.A. Williams
1991 *Introduction to Statistics* St. Paul: West Publishing Company.

Baayen, R. Harald
2001 *Word Frequency Distributions.* Dordrecht, Boston, London: Kluwer.

Biskin, Bruce H.
1980 Multivariate analysis in experimental counseling research. *The Counseling Psychologist* 8(4), 69-73.

Box, G.E.P.
1954 Some theorems on quadratic forms applied in the study of variance problems, II: Effect of inequality of variance and of correlation between errors in the two-way classification. *Annals of Mathematical Statistics* 25, 484–498.

Bryk, Anthony S. and Stephen W. Raudenbush
1988 Heterogeneity of variance in experimental studies: A challenge to conventional interpretations. *Psychological Bulletin* 104, 396–404.

Butler, C.
1985 *Statistics in Linguistics.* Oxford: Basil Blackwell

Chen, Aoju, Carlos Gussenhoven and Toni Rietveld
2004 Language-specificity in the perception of paralinguistic intonational meaning. *Language and Speech* 47: 311–349.

Clark, Herbert H.
1973 The language-as-fixed-effect fallacy: A critique of language Statistics in Psychological Research. *Journal of Verbal Learning and Verbal Behavior* 12, 335–359.

Cohen, Jacob and P. Cohen
1985 *Applied Multiple Regression and Correlation Analysis for the Behavioral Sciences.* (2nd ed.) Hillsdale, NJ: Erlbaum.

Cohen, Jacob
1988 *Statistical Power Analysis for the Behavioral Sciences. (2nd ed.)* Hillsdale, NJ: Erlbaum.

Coleman, Edmund B.
1979 Generalization effects vs random effects: Is σ^2_{TL} a source of type 1 or type 2 error? *Journal of Verbal Learning and Verbal Behavior* 18: 243–256.

Cornfield, Jerome and John W. Tukey
1956 Average Values of Mean Squares in Factorials. *The Annals of Mathematical Statistics* 27, 907–949.

Efron, B. and G. Gong
1983 A leisurely look at the bootstrap, the jackknife, and cross-validation. *The American Statistician* 37, 26–49.

Forster, K.I. and R.G. Dickinson
1976 More on the language-as-fixed-effect fallacy: Monte Carlo estimates of error rates for F1, F2, F' and min F'. *Journal of Verbal Learning and Verbal Behavior* 15, 135–142.

Glantz, Stanton A. and B.K. Slinker
1990 *Primer of Applied Regression and Analysis of Variance.* New York: McGraw-Hill, Inc.

Gornbein, Jeffrey G., Carlos Lazaro and Roderick J.A. Little
1992 Incomplete data in repeated measures analysis. *Statistical Methods in Medical Research* 1, 275–295.

Harris, Richard J.
1975 *A Primer of Multivariate Statistics.* New York: Academic Press.

Hays, William L.
1981 *Statistics.* New York: Holt, Rinehart and Winston.

Huynh, Huynh S. and Leonard S. Feldt
1976 Estimation of the box correction for degrees of freedom from sample data in randomized block and split plot designs. *Journal of Educational Statistics* 1, 69–82.

Kenny, David A.
1990 Design issues in dyadic research. In: C. Hendrick and M.S. Clark (eds.), *Research Methods in Personality and Social Psychology* Beverly Hills: Sage Publications, 164–184.

Kenny, David A. and Charles M. Judd
1986 Consequences of violating the independence assumption in analysis of variance. *Psychological Bulletin* 99, 422–431.

Kirk, Roger E.
1995 *Experimental Designs: Procedures for the Behavioral Sciences.* (3rd ed.) Belmont: Brooks/Cole.

Labovitz, Sanford
1970 The assignment of numbers to rank order categories. *American Sociological Review* 35, 515–524.

Little, Roderick J.A. and D.A. Rubin
2002 *Statistical Analysis with Missing Data.* (2nd ed.) New York: John Wiley.

Little, Roderick J.A. and N. Schenker
1995 Missing data. In: *Handbook of Statistical Modeling for the Social and Behavioral Sciences* New York: Plenum.

Lunneborg, Clifford E.
1987 *Bootstrap Applications for the Behavioral Sciences, Vol. 1.* Washington: Dept. of Psychology, University of Washington.

Lunney, Gerald H.
1970 Using analysis of variance with a dichotomous dependent variable: an empirical study. *Journal of Education Measurement* 7, 263–269.

Maxwell, Scott E, and Bray, James H.
1986 Robustness of the Quasi F Statistic to Violations of Sphericity. *Psychological Bulletin* 99, 416–421.

McCulloch, Charles E. and S.R. Searle
2001 *Generalized, Linear, and Mixed Models.* New York: John Wiley and Sons, Inc.

Milliken, George and Dallas E. Johnson
1997 *Analysis of Messy Data. Vol. 1: Designed Experiments.* London: Chapman and Hall.

Milroy, Lesley
1980 *Language and Social Networks.* Oxford: Basil Blackwell.

Nolan, F.
2003 Intonational equivalence: an experimental evaluation of pitch scales. *Proceedings 15th International Congress of Phonetic Sciences, Barcelona.* 771–774.

Oakes, Michael P.
1998 *Statistics for Corpus Linguistics.* Edinburgh: Edinburgh University Press.

Page, Melanie C., Sanford L. Braver and David P. MacKinnon
2003 *Levine's Guide to SPSS for Analysis of Variance.* (2nd ed.) London: Lawrence Erlbaum Associates, Publishers.

Quene, Hugo and Huub van den Bergh
2004 On multi-level modeling of data from repeated measures designs: A tutorial. *Speech Communication* 43, 103–121.

Richards, Brian
1987 Type/Token ratios: What do they really tell us? *Journal of Child Language* 14, 201–209.

Rietveld, Toni and Roeland van Hout
1993 *Statistical Techniques for the Study of Language and Language Behavior.* Berlin, New York: Mouton de Gruyter.

Rietveld, Toni, Joop Kerkhoff and Carlos Gussenhoven
2004 Word prosodic structure and vowel duration in Dutch. *Journal of Phonetics* 32, 349–371.

Rogan, J.C., H.J. Keselman and J.L. Mendoza
1979 Analysis of Repeated Measurements. *British Journal of Mathematical and Statistical Psychology* 32, 269–286.

Rubin, Donald B.
1976 Inference and missing data. *Biometrika* 63: 581–592.

Rubin, Donald B.
1987 *Multiple Imputation for Nonresponse in Surveys.* New York: Wiley.

Rubin, Donald B.
1991 *Multiple Imputation for Nonresponse in Surveys.* New York: Wiley.

Santa, J.L., J.J. Miller and M.L. Shaw
1979 Quasi F to Prevent Alpha Inflation Due to Stimulus Variation. *Psychological Bulletin* 86, 37–46.

Siegel, Sidney and N. John Jr. Castellan
1988 *Nonparametric Statistics for the Behavioral Sciences.* (2nd ed.) New York: McGraw-Hill.

Stevens, James
1986 *Applied Multivariate Statistics for the Social Sciences.* Hillsdale, New Jersey: Lawrence Erlbaum.

Stevens, S. S.
1946 On the theory of scales of measurement. *Science* 103, 677-680.

Stevens, S.S.
1951 Mathematics, measurement, and psychophysics. In: S.S. Stevens (ed.), *Handbook of Experimental Psychology.* New York, 1-49.

Studebaker, Gerald A.
1985 A 'rationalized' arcsine transformation. *Journal of Speech and Hearing Research* 28, 455–462.

Tatsuoka, M.M.
1971 *Multivariate Analysis.* New York: Wiley.

Todman, John B. and Pat Dugard
2001 *Single-case and Small-n Experimental Designs.* London: Lawrence Erlbaum Associates, Publishers.

Welch, B.L.
1947 The generalization of student's problem when several different population variances are involved. *Biometrika* 3628, 293–296.

Westfall, P.H.
1987 A comparison of variance components estimates for arbitrary underlying distributions. *Journal of the American Statistical Association* 82, 866-874.

Wilcox, Rand R.
1987 New designs in analysis of variance. *Annual Review of Psychology* 38, 29–60.

Winer, B.J., D.R. Brown and K.M. Michels
1991 *Statistical Principles in Experimental Design.* New York: McGraw-Hill.

Woods, A., P. Fletcher and A. Hughes
1986 *Statistics in Language Studies.* Cambridge: Cambridge University Press.

Appendix A: Key to the exercises

Chapter 2: Basic statistical procedures: one sample

1. $z = 2.80$; a) one-tailed: yes, b) two-tailed: yes.
   $z = 1.65$; a) one-tailed: yes, b) two-tailed: no.
   $z = -1.75$; a) one-tailed: yes, b) two-tailed: no.
2. If a clear alternative hypothesis pointing in one direction exists (for instance A $>$ B) a one-tailed test should be carried out. If the alternative hypothesis is just A $\neq$ B, a two-tailed test is appropriate.
3. Larger samples do increase the power of a test, as the *SE* decreases with increasing sample size (= *N*).
4. As the power of a test = 1 – (probability of committing a Type II error), the increase of a Type II error will decrease the power.
5. $t_{34} = \frac{(505-500)/(20}{\sqrt{35}} = 1.479$, ns (= not significant).
6. Using SAMPLEPOWER or by doing the calculations by hand we obtain a power of 61% when we compare a proportion of 65% with a proportion of 50%.
7. SAMPLEPOWER tells us that the sample size (written as *N* or *n*) needed to achieve the desired power of 80% is 154.
8. $p = 0.05$,

Chapter 3: Basic statistical procedures: two samples

1. $t_4 = 2.80$; a) one-tailed: yes, b) two-tailed: yes.
   $t_{65} = 1.65$; a) one-tailed: no, b) two-tailed: no.
   $t_{14} = -1.75$; a) one-tailed: yes, b) two-tailed: no.
2. a) Independent samples: $t_8 = 0.203$, ns.
   b) Paired samples: $t_4 = -2.631$, significant at the 5% level.
   The correlation between the scores in the two conditions is considerable: 0.98, and has consequently a large influence on the standard error (see equation (5) in Chapter 3).
3. The z test for comparing proportions yields, on the basis of the standard error calculated by $\sqrt{\frac{0.437\times0.563}{16} + \frac{0.733\times0.267}{30}} = 0.148$, a value of $\frac{-0.296}{0.148} =$ –2.00; significant at the 5% level (one-tailed).

4. Using the program SAMPLEPOWER we find n_1 and $n_2 = 19$; 38 children in total.
5. There are no populations to which a generalization of the means of the chronological ages is relevant ('school classes 3A and 3B'). That is why a comparison of the ages with a *t* test is pointless.

Chapter 4: Analysis of variance

1. $F_{10,20} = 2.80$; yes.
   $F_{2,8} = 3.10$; no.
   $F_{2,80} = 3.10$; no.
   $F_{2,800} = 3.10$; yes.
   $F_{5,100} = 2.25$; no.
   $F_{10,100} = 2.25$; yes.
2. The complete table is:

Source	*SS*	*df*	*MS*	*F*
Between	200	4	50.0	4
Within	500	40	12.5	
Total	700	44		

3. The resulting ANOVA table is:

Source	*SS*	*df*	*MS*	*F*
Between	54.0	2	27.0	27.0
Within	6.0	6	1.0	
Total	60.0	8		

4. $$\begin{aligned} SS_{\text{between}} &= 4 \times (3.75 - 5.25)^2 + 4 \times (7.00 - 5.25)^2 + 4 \times (5.00 - 5.25)^2 \\ &= 21.500 \\ MS_{\text{between}} &= 21.500/2 = 10.750 \\ SS_{\text{within}} &= (5 - 3.75)^2 + (2 - 3.75)^2 + (4 - 3.75)^2 + (4 - 3.75)^2 + \\ &\quad (7 - 7)^2 + (6 - 7)^2 + (7 - 7)^2 + (8 - 7)^2 + \\ &\quad (5 - 5)^2 + (4 - 5)^2 + (6 - 5)^2 + (5 - 5)^2 \\ &= 8.750 \\ MS_{\text{within}} &= 8.750/9 = 0.972 \end{aligned}$$
5. The maximum number of homogeneous sets is 6: ((1+2), 3), ((1+3), 2), ((2+3),1); every level of the factor 'Treatment' can also be a 'homogeneous set' of its own.

6. a) μ
   b) $\mu + \varepsilon$
7. No, only *SS*s can be added.
8. $df_1 = 4$, $df_2 = 65$.

Chapter 5: Multifactorial designs

1. Without interaction the cell means would be, on the basis of estimates of the effects α_i and β_j (cf. $\alpha_i = \mu_{i.} - \mu$ and $\beta_j = \mu_{.j} - \mu$):

	A_1	A_2	A_3
B_1	2	8	14
B_2	10	16	22

2. This is a one-way design, with just one factor: group of speakers. 'Nesting' and 'crossing' are no relevant terms here.
3. $F = 56.333/1.333 = 42.260$.
4. $X_{ijk} = \mu + \alpha_i + \beta_{j(i)} + \varepsilon_{ijk}$
5. In (2), as a nested factor cannot show an interaction with the factor within it is nested.
6. Yes, see Figure 5.3.
7. a) Factor A: $F_{1.8} = 0.062$, $p = 0.809$; Factor B: $F_{1.8} = 10.563$, $p = 0.012$; the interaction AB: $F_{1,8} = 0.062$, $p = 0.809$.
   b) Before the analysis: $X_{ijk} = \mu + \alpha_i + \beta_j + \alpha\beta_{i,j} + \varepsilon_{ijk}$. After the analysis: $X_{ijk} = \mu + \beta_j + \varepsilon_{jk}$ as neither the factor A nor the interaction AB are significant.
8. a) Factor A: $F_{1.4} = 54.760$, $p = 0.002$; Factor B: $F_{4.18} = 0.497$, $p = 0.738$.
   b) The syntax is:

```
UNIANOVA dep BY a b
  /RANDOM = b
  /METHOD = SSTYPE(3)
  /INTERCEPT = INCLUDE
  /CRITERIA = ALPHA(.05)
  /DESIGN = a, b WITHIN a.
```

9. a) Therapy: $F_{2.26} = 1.812, p = 0.183$, ns.
   b) The interaction 'therapy by T' is not significant: $F_{2.24} = 2.70, p = 0.087$, ns. The assumption of equal variances is warranted.

Chapter 6: Additional tests and indices

1. A: $\eta^2 = SS_A/SS_{\text{corrected total}} = 5.333/93.667 = 0.057$
   B: $\eta^2 = SS_B/SS_{\text{corrected total}} = 48.000/93.667 = 0.512$
   AB: $\eta^2 = SS_{AB}/SS_{\text{corrected total}} = 3.000/93.667 = 0.032$
   A: $\eta^2_{\text{partial}} = 5.333/(5.333{+}37.333) = 0.125$
   B: $\eta^2_{\text{partial}} = 48.000/(48.000{+}37.333) = 0.563$
   AB: $\eta^2_{\text{partial}} = 3.000/(3.000{+}37.333) = 0.074$
2. B within A(1): $F_{1,9} = 2.95$; $p = 0.126$
   B within A(2): $F_{1,9} = 7.91$; $p = 0.020$
3. Assumed equal variances: $t_{15} = -4.119$; $p = 0.001$.
4. The calculation of ω^2 is based on estimates of variance components in the population; η^2 is only based on values of *SS*.
5. There are $k - 1$ orthogonal contrasts possible; in this example $5 - 1 = 4$.
6. a) R^2 is the *SS* associated with the Corrected Model divided by the *SS* of the Corrected Total: $\frac{627.000}{1108.500} = 0.566$. $R^2_{\text{adjusted}} = \frac{0.566-(1-0.566)\times 5}{(24-5-1)} = 0.443$.
   b) $\eta^2_{\text{rate}} = 322.667/1108.50 = 0.29$; $\eta^2_{\text{dialect}} = 84.250/1108.50 = 0.08$; $\eta^2_{\text{rate} \times \text{dialect}} = 220.083/1108.50 = 0.20$

Chapter 7: Violations of assumptions and unbalanced designs

1. Levene's statistic: $F_{2,9} = 6.313$, $p = 0.019$
2. For the transformation $2\arcsin\sqrt{p}$ the ratio is: 0.217/0.039 = 7.233; for Studebaker's transformation the ratio is: 0.213/0.029 = 7.345.
3. a) Levene's $F_{2,12} = 6.829$, $p = 0.010$
   b) Oneway: $F_{2,12} = 9.487$, $p = 0.003$
   c) Welch's test: $F_{2,5.803} = 31.315$, $p = 0.001$
   d) A ceiling effect occurred: either the children assigned to Method 2 were very intelligent, or Method 2 leads to outstanding results.
4. When x can take the value of 0.
5. $\beta_1 - \beta_2$ is estimated by: $\frac{1}{4} \times [(\alpha_1 + \beta_1) + (\alpha_1 + \beta_1) + (\alpha_2 + \beta_1) + (\alpha_2 + \beta_1)] - \frac{1}{3} \times [(\alpha_1 + \beta_2) + (\alpha_1 + \beta_2) + (\alpha_2 + \beta_2)] = -\frac{1}{6}\alpha_1 + \frac{1}{6}\alpha_2 + \beta_1 - \beta_2$
6.

Source	*SS*	*df*	*MS*	*F*	*p*
gender	2.155	2	2.155	11.994	.001
network	6.09	1	0.609	3.389	.075
gender × network	0.211	1	1.176	.286	

Chapter 8: Repeated measures designs

1. The design is:

	CV		CVC		('target')
	CV	CVC	CV	CVC	('syllable boundary')
	w1 ... w10	w11 ... w20	w1 ... w10	w11 ... w20	('word')
1					
⋮					
20					
('Subjects')					

Crossed: listeners ('subjects') with all factors, 'syllable boundary' with 'target', 'word' with 'target'.
Nested: 'word' within 'word type'.

2. $F1$: $MS_{\text{stress}} = 4305.625$, $MS_{\text{error}} = 272.812$; $F1_{1,4} = 15.78$, $p = 0.017$; $F2$: $\bar{X}_{w1} = 504, \bar{X}_{w2} = 497, \bar{X}_{w3} = 465, \bar{X}_{w4} = 453$. $MS_{\text{accent}} = 1722.250$, $MS_{\text{error}} = 48.250$; $F1_{1,2} = 35.694$, $p = 0.027$. $F1 \times F2$ criterion: significant. $F'(1,6) = 11.13$, $p < 0.05$ (df rounded from 1.02 and 5.98 respectively); $\text{min}F'(1,6) = 11.242$, $p < 0.05$.
3. Larger p values.
4. H-F should be applied to the df associated with the within-subject factor.
5. $F_{4.16} = 2.20$; $p > 0.05$.
6. a) CV-type: $F = MS_{\text{CV-type}}/MS_{\text{CV-type} \times \text{babies}}$
   b) babies: $F = MS_{\text{babies}}/MS_{\text{within}}$.
7. No, as no MS_{within} is available.

Chapter 9: Alternative estimation procedures and missing data

1. In the ANOVA approach the factor 'group' is not significant at the 5% level: $F_{2.9} = 3.566$, $p = 0.072$. In the ML approach this factor is significant: $F_{2.12} = 4.755$, $p = 0.030$. The difference is the result of two things: a) different degrees of freedom associated with the error term (9 vs. 12), and b) different estimates of the error. For ANOVA it is 5.889, for the ML approach it is 4.416.
2. As $(1 - 1/k)MS_A \geq MS_E (20.321 \geq 4.667)$, the formula $\hat{\sigma}^2_\alpha = \frac{(1-1/k)MS_A - MS_E}{N}$ is used: $\frac{(1-1/3)30.467-4.667}{15} = 1.044$
3. For the oneway design AIC = 59.879. Including the factor 'Factor2' de-

creases the value of AIC; it becomes 57.458; the value of -2 Log Likelihood drops from 51.879 to 47.458.

4. a) Complete data set, factor 'type': $F_{1,3} = 316.000$, $p = 0.000$.
   b) The analysis of variance carried out on the data set with the two missing data does not result in an F ratio for the factor 'type', as the df associated with the term 'residual + interaction' is too small. MVA, option REGRESSION imputes 590 for missing value (s1,t1,w1) and 630 for (s3,t2,w3). The analysis of variance carried out on the data set with two imputed values yields for factor Type: $F_{1,3} = 150.000$, $p = 0.001$.

Chapter 10: Alternatives to analysis of variance

1. Six arrangements are possible. The values of $|D|$ - the differences between the means obtained in C1 and C2 are 11 (the original data set) and 3, 2, 11, 3 and 2. A value of $|D| \leq 11$ occurred in 2 of the six arrangements, so p is 2/6 = 0.333, two-tailed: not significant at the 5% level.
2. In bootstrap procedures.
3. No. Both the F ratio and SS_{error} are indices of the ratio of MS_{between} and MS_{within}. As the theoretical distribution of the F ratio is not used both indices are equally valid for use in a resampling procedure.
4. In $\frac{10!}{5! \times 5!} = \frac{3628800}{14400} = 252$.
5. No, it is the user of the bootstrap procedure who determines the number of bootstraps with replacement; in principle this number is infinite as the used data are replaced.
6. $Y_{ij} = \beta_{0j} + \beta_{1j} \times TYPE + \varepsilon$ is a level-1 equation, both $\beta_{0j} = \gamma_{00} + \gamma_{01} \times MT + u_{0j}$ and $\beta_{1j} = \gamma_{10} + \gamma_{11} \times MT + u_{1j}$ are level-2 equations.

Appendix B: Matrix algebra

Matrix algebra is a branch of mathematics with a wide range of applications. It is a basic tool in multivariate statistics and its concise notational system is most useful for analysing complex research data and conceptualizing problems in empirical linguistic research. This appendix only provides a global introduction to some key concepts from matrix algebra.

Matrices and vectors

In many software packages for data analysis it is assumed that the data are organized in such a way that the rows (horizontal arrays) represent the participants (cases) and the columns (vertical arrays) the variables on which the participants have been measured (see Chapter 1). If we only have three participants and one variable, the data can be described in the following format:

participant 1:	3.0
participant 2:	1.2
participant 3:	5.0

An alternative way of presenting these data is the following notation in which a pair of brackets is used:

$$\begin{pmatrix} 3.0 \\ 1.2 \\ 5.0 \end{pmatrix}$$

The values or scores of the variable can even be described in a geometric way by putting the scores on a straight line.

Now suppose that we have observed two variables. Can they be described in a similar way? It is not difficult to see that the following notations (for two variables; the bivariate case) are a straightforward extension of the notation used for one variable (the univariate case):

participant 1:	3.0	4.0
participant 2:	1.2	2.4
participant 3:	5.0	2.5

$$\begin{pmatrix} 3.0 & 4.0 \\ 1.2 & 2.4 \\ 5.0 & 2.5 \end{pmatrix}$$

The notation between brackets is typical of matrix notation: a matrix is an array of numbers arranged in rows and columns. The individual entries of a matrix are usually called elements. Normally, the elements of each row or column have something in common. In our example, for instance, the third row represents the scores of participant 3; the second column gives the scores on the second variable. Thus, a matrix can be defined as a rectangular array of $m \times n$ numbers arranged in m rows and n columns (an m by n matrix). In the preceding example $m = 3$ (the number of rows) and $n = 2$ (the number of columns). The matrix notation can also be used for more than two variables as well as for a larger number of participants, simply by expanding the number of columns and rows respectively.

Matrices are commonly symbolized by capitals. An extended notation for the data presented above is:

$$A = \begin{pmatrix} 3.0 & 4.0 \\ 1.2 & 2.4 \\ 5.0 & 2.5 \end{pmatrix}$$

A is a $3{\times}2$ matrix: 3 rows and 2 columns. Each row or column is a vector. The pair (1.2, 2.4) is a 2-component row vector. The set:

$$\begin{pmatrix} 4.0 \\ 2.4 \\ 2.5 \end{pmatrix}$$

is a 3-component column vector. Vectors are symbolized by small letters, e.g., a. The numbers or values in a matrix are its elements. Element a_{21} refers to the element in the second row and the first column. This element has the value 1.2 in the preceding matrix A. Matrix A can also be written as $\{a_{ij}\}$.

Matrices frequently occur in statistical analysis. As shown by the examples, the observed data are first put in the form of a matrix. Subsequent algebraic operations can be carried out on such a matrix in order to get matrices relevant to statistical analysis. This is shown in the final section of this appendix by calculating a correlation and a covariance matrix.

Matrix operations

The kinds of calculations or operations which can be carried out with vectors and matrices are the ones familiar from standard algebraic calculus, e.g., addition, subtraction and multiplication. These operations are defined in a way comparable to the 'normal' definitions known from standard algebra, but they are adapted to the situation of handling matrices. Instead of giving a strictly mathematical treatment of matrix algebra, we will only demonstrate some matrix operations on the basis of simple examples.

addition of matrices: $A + B = \{a_{ij} + b_{ij}\} = \{c_{ij}\} = C$

$$\begin{pmatrix} 1 & 4 \\ 5 & 3 \\ 6 & 2 \end{pmatrix} + \begin{pmatrix} 2 & 3 \\ 6 & 3 \\ 5 & 8 \end{pmatrix} = \begin{pmatrix} 3 & 7 \\ 11 & 6 \\ 11 & 10 \end{pmatrix}$$

The sum of two matrices A and B is a matrix C with elements c_{ij}, which are obtained by adding the corresponding elements of A and B, e.g., $a_{11} + b_{11} = c_{11}$, $a_{32} + b_{32} = c_{32}$. Since only corresponding elements can be added, the sum of two matrices is only defined when both matrices have the same size. Matrices are said to have the same size or *same order* when they have the same number of rows and columns.

subtraction of matrices: $A - B = \{a_{ij} - b_{ij}\} = \{c_{ij}\} = C$

Subtraction of matrices proceeds in a way similar to addition, the only difference being that corresponding elements are not added but subtracted.

multiplication of a matrix by a scalar: $\alpha \times A = \{\alpha a_{ij}\} = \{c_{ij}\} = C$

$$2 \times \begin{pmatrix} 1 & 4 \\ 5 & 3 \\ 6 & 2 \end{pmatrix} = \begin{pmatrix} 2 & 8 \\ 10 & 6 \\ 12 & 4 \end{pmatrix}$$

A scalar is a single number. It is symbolized by a letter from the Greek alphabet. The multiplication of a matrix (A) by a scalar (α) is obtained by multiplying each element of the matrix by this scalar. The resulting matrix (C) is of the same order as the original matrix (A). The operation is simple and it is obvious that every matrix can be multiplied by a scalar.

the product of a row vector and a column vector:

$$a \times b = \{\sum_{k=1}^{n} a_k \times b_k\} = \alpha$$

$$(1\ 5\ 6) \times \begin{pmatrix} 1 \\ 2 \\ 4 \end{pmatrix} = (1 \times 1 + 5 \times 2 + 6 \times 4) = 35$$

The first step in getting the product of a row and a column vector is the multiplication of corresponding row and column elements ($a_1 \times b_1$, $a_2 \times b_2$, etc.), followed by the summing of the products. The result is a single number, a scalar. This is why the product of a row and a column vector is referred to as a *scalar product*. As corresponding elements are multiplied, the number of elements in the row and the column vector must be equal, otherwise multiplication is not possible. Scalar products of vectors frequently occur in multivariate analysis. Many concepts familiar from statistical analysis can be defined as scalar products. The sum of the observed values of a variable (in summation notation $\sum X$), for example, can be defined as the sum of the elements of a column vector (a) (pre)multiplied by a unit row vector (j; all elements in this vector have the value 1) with the same dimensionality (both vectors have the same number of elements or components). Thus:

$$\sum X \Rightarrow j \times a = ja = (1\ 1\ 1\ 1\ 1) \times \begin{pmatrix} 1 \\ 2 \\ 4 \\ 2 \\ 7 \end{pmatrix} = 16$$

The sum of the product of the observed values of two variables, X and Y (in summation notation $\sum XY$), can be obtained by multiplying the values of X as a row vector (a) by the values of Y as a column vector (b):

$$\sum XY \Rightarrow ab = (1\ 2\ 4\ 2\ 7) \times \begin{pmatrix} 3 \\ -5 \\ 1 \\ 3 \\ -4 \end{pmatrix} = -25$$

the product of two matrices:

$$A \times B = \{\sum_{k=1}^{n} a_{ik} \times b_{kj}\} = \{c_{ij}\} = C$$

$$\begin{pmatrix} 1 & 4 \\ 5 & 3 \\ 6 & 2 \end{pmatrix} \times \begin{pmatrix} 1 & 2 \\ 7 & 4 \end{pmatrix} = \begin{pmatrix} 1 \times 1 + 4 \times 7 & 1 \times 2 + 4 \times 4 \\ 5 \times 1 + 3 \times 7 & 5 \times 2 + 3 \times 4 \\ 6 \times 1 + 2 \times 7 & 6 \times 2 + 2 \times 4 \end{pmatrix} = \begin{pmatrix} 29 & 18 \\ 26 & 22 \\ 20 & 20 \end{pmatrix}$$

Multiplying two matrices is a more complicated operation than the ones described before. The multiplication of matrices is a straightforward extension of the operation of multiplying a row vector by a column vector. The elements in the resulting matrix C are obtained by multiplying every row vector of matrix A by every column vector of matrix B. For example, the element c_{11} is calculated as follows:

$$c_{11} = \sum_{k=1}^{n} a_{1k} \times b_{k1} = (1\ 4)\begin{pmatrix} 1 \\ 7 \end{pmatrix} = a_{11} \times b_{11} + a_{12} \times b_{21} = 1 \times 1 + 4 \times 7 = 29$$

The other elements are obtained in a similar way:

$$\begin{aligned}
c_{12} &= a_{11} \times b_{12} + a_{12} \times b_{22} = 1 \times 2 + 4 \times 4 = 18 \\
c_{21} &= a_{21} \times b_{11} + a_{22} \times b_{21} = 5 \times 1 + 3 \times 7 = 26 \\
c_{22} &= a_{21} \times b_{12} + a_{22} \times b_{22} = 5 \times 2 + 3 \times 4 = 22 \\
c_{31} &= a_{31} \times b_{11} + a_{32} \times b_{21} = 6 \times 1 + 2 \times 7 = 20 \\
c_{32} &= a_{31} \times b_{12} + a_{32} \times b_{22} = 6 \times 2 + 2 \times 4 = 20
\end{aligned}$$

From the definition of the multiplication operation it becomes clear that not every pair of matrices can be multiplied. Two matrices can only be multiplied if the number of columns of the first matrix is equal to the number of rows of the second. If A is an $n \times p$ matrix, it can only be multiplied by a matrix B which is a $p \times m$ matrix. The product C is an $n \times m$ matrix. The definition of the product also implies that $A \times B$ is not equal to $B \times A$. In our example it is not possible to calculate the product $B \times A$: the number of columns of the first matrix (B has 2 columns) is not equal to the number of rows of the second matrix (A has 3 rows).

Some key concepts in matrix algebra

The transpose

In several statistical operations where matrices are used, the transpose of a matrix plays a crucial role. The transpose is obtained by interchanging the rows and columns of a matrix. Let $A = \{a_{ij}\}$ be an $m \times n$ matrix. The transpose, written A' or A^t, therefore is the $n \times m$ matrix obtained by interchanging the rows and columns of A. The matrix A' can be written as $\{a_{ji}\}$. For example:

$$A = \begin{pmatrix} 1 & 4 \\ 5 & 3 \\ 6 & 2 \end{pmatrix} \quad \text{then } A' = \begin{pmatrix} 1 & 5 & 6 \\ 4 & 3 & 2 \end{pmatrix}$$

Because of the interchange of rows and columns, the number of rows of a matrix A always equals the number of columns of the transpose A'. That is why it is always possible to multiply A' by A (and also A by A'; but the outcomes are not the same). The product matrix $S = A' \times A$ has some useful properties for statistical calculations.

$$S = A'A = \begin{pmatrix} 1 & 5 & 6 \\ 4 & 3 & 2 \end{pmatrix} \times \begin{pmatrix} 1 & 4 \\ 5 & 3 \\ 6 & 2 \end{pmatrix} = \begin{pmatrix} 62 & 31 \\ 31 & 29 \end{pmatrix}$$

The elements on the diagonal of the product matrix S are composed of the squared values of the elements in the columns of matrix A:

$$s_{11} = a_{11}^2 + a_{21}^2 + a_{31}^2 = 1^2 + 5^2 + 6^2 = 62$$
$$s_{22} = a_{12}^2 + a_{22}^2 + a_{32}^2 = 4^2 + 3^2 + 2^2 = 29$$

The sum of squares of the values of a variable ($\sum X^2$) has an important function in many statistical formulas; the formula of the sample variance is a good example. The other two elements in matrix S can be interpreted as the sum of the product of two different variables ($\sum XY$) which plays a crucial role in the definition of covariance and correlation.

The determinant of a matrix

The determinant is a scalar function of the matrix elements or entries. Remember that a scalar is a single number. Only square matrices have a determinant. The calculation of determinants of large matrices is relatively tedious (cf. Grossman 1978:91ff), but for a 2×2 matrix it is quite simple. The determinant is the product of the elements on the diagonal minus the product of the two remaining elements:

$$\det A = |A| = \begin{vmatrix} a_{11} & a_{12} \\ a_{21} & a_{22} \end{vmatrix} = a_{11}a_{22} - a_{12}a_{21}$$

The determinant of a matrix A is denoted as $|A|$. In statistical analysis, correlation and covariance matrices are square matrices and their determinants convey important information. The determinant is, for instance, related to measures of variance (for instance, the amount of explained variance).

When we think in terms of column vectors, we say that a column vector is linearly dependent on one or more other column vectors if it can be rewritten as a linear function of these vectors. An example illustrates this definition. Suppose we have the following two vectors, v_1 and v_2:

$$\begin{pmatrix} 4 \\ 2 \end{pmatrix} \quad \begin{pmatrix} 8 \\ 4 \end{pmatrix}$$

The second vector can be rewritten as a linear combination of the first one and a scalar: $v_2 = 2 \times v_1$. In matrix notation we get the following result:

$$\begin{pmatrix} 8 \\ 4 \end{pmatrix} = 2 \times \begin{pmatrix} 4 \\ 2 \end{pmatrix}$$

We may conclude that both vectors are linearly dependent. When we have a matrix V containing both vectors, we get:

$$V = \begin{pmatrix} 4 & 8 \\ 2 & 4 \end{pmatrix}$$

The determinant of this matrix is zero:

$$|V| = \begin{vmatrix} 4 & 8 \\ 2 & 4 \end{vmatrix} = 4 \times 4 - 8 \times 2 = 0$$

It is always the case that when rows or columns of a matrix are linearly dependent, the determinant of that matrix is zero. A matrix with a determinant of zero is called *singular*. If the determinant is not equal to zero, the matrix is said to be *nonsingular*.

The inverse of a matrix

The inverse of a (square) matrix A is notated as A^{-1}. The multiplication of A and A^{-1} is equal to I, the identity matrix. For example:

$$A \times A^{-1} = \begin{pmatrix} 1 & 2 \\ 3 & 4 \end{pmatrix} \times \begin{pmatrix} -2 & 1 \\ 3/2 & -1/2 \end{pmatrix} = \begin{pmatrix} 1 & 0 \\ 0 & 1 \end{pmatrix} = I$$

In fact, such a matrix product is equivalent to a scalar product like $5 \times 5^{-1} = 5 \times \frac{1}{5} = 1$. Not all matrices have an inverse. It is important to note that a singular matrix (whose determinant is zero) has no inverse. Furthermore, multiplication by the inverse in matrix algebra is comparable to dividing in ordinary algebra.

Some matrix operations for statistical data

The measurements or scores of three participants on two variables can be written in the form of a data matrix, e.g.:

$$X = \begin{pmatrix} 12 & 22 \\ 8 & 13 \\ 10 & 16 \end{pmatrix}$$

The compactness of the matrix notation of statistical measures can be illustrated by computing the following four matrices:

- raw cross-product matrix	$B = X'X$
- SSCP matrix (sums of squares and cross products)	$S = X'X - \frac{1}{N}(X'1)(1'X) = X_d'X_d$
- covariance matrix	$C = \frac{1}{N-1}S$
- correlation matrix	$R = DSD$

The raw cross-product matrix is computed by multiplying matrix X by its transpose:

$$B = X'X \quad = \quad \begin{pmatrix} 12 & 8 & 10 \\ 22 & 13 & 16 \end{pmatrix} \times \begin{pmatrix} 12 & 22 \\ 8 & 13 \\ 10 & 16 \end{pmatrix} = \begin{pmatrix} 308 & 528 \\ 528 & 909 \end{pmatrix}$$

B is a symmetric matrix. In the more familiar notation with the sum sign, the four elements in matrix B are equal to:

$$\begin{matrix} \Sigma X_1^2 & \Sigma X_1 X_2 \\ \Sigma X_1 X_2 & \Sigma X_2^2 \end{matrix}$$

This matrix B forms part of the formula for the SSCP matrix S. In the formula for S the part on the right accounts for the fact that the raw cross-products have to be replaced by the sums of squares of deviations. In the summation notation the elements of the matrix should have the value:

$$\sum x_i x_j \quad = \quad \sum X_i X_j - \frac{(\Sigma X_i \Sigma X_j)}{N}$$

where N denotes the number of observations on which the variances and covariances are based.

In the formula for S, '1' denotes an $n \times 1$ unit vector. $X'1$ and $1'X$ are:

$$X'1 = \begin{pmatrix} 12 & 8 & 10 \\ 22 & 13 & 16 \end{pmatrix} \begin{pmatrix} 1 \\ 1 \\ 1 \end{pmatrix} = \begin{pmatrix} 30 \\ 51 \end{pmatrix}$$

$$1'X = (1\ 1\ 1)\begin{pmatrix} 12 & 22 \\ 8 & 13 \\ 10 & 16 \end{pmatrix} = (30\ 51)$$

The values 30 and 51 represent $\sum X_1$ and $\sum X_2$ respectively. The right-hand part of the formula for S can now be calculated as follows:

$$\frac{1}{N}(X'1)(1'X) = \frac{1}{N}\begin{pmatrix} 30 \\ 51 \end{pmatrix}(30\ 51) = \begin{pmatrix} 300 & 510 \\ 510 & 867 \end{pmatrix}$$

The four elements in this matrix can be represented in summation notation:

$$\begin{matrix} \frac{(\Sigma X_1)^2}{N} & \frac{\Sigma X_1 \Sigma X_2}{N} \\ \frac{\Sigma X_1 \Sigma X_2}{N} & \frac{(\Sigma X_2)^2}{N} \end{matrix}$$

Matrix S (the SSCP matrix) is:

$$S = \begin{pmatrix} 308 & 528 \\ 528 & 909 \end{pmatrix} - \begin{pmatrix} 300 & 510 \\ 510 & 867 \end{pmatrix} = \begin{pmatrix} 8 & 18 \\ 18 & 42 \end{pmatrix}$$

The elements of matrix S can again be represented in summation notation:

$$\begin{matrix} \Sigma x_1^2 & \Sigma x_1 x_2 \\ \Sigma x_1 x_2 & \Sigma x_2^2 \end{matrix}$$

An alternative to this method of calculating matrix S is based on the deviation scores x_i. The mean value of the first variable ($= X_1$) is 10; the mean value of the second variable (X_2) is 17. By subtracting $(X - \bar{X})$ the deviation scores (x_i) can be computed. The resulting matrix is:

$$X_d = \{x\} = \begin{pmatrix} 2 & 5 \\ -2 & -4 \\ 0 & -1 \end{pmatrix}$$

If this matrix X_d is multiplied by its transpose, we get the following result:

$$X_d'X_d = \{x'x\} = \begin{pmatrix} 2 & -2 & 0 \\ 5 & -4 & -1 \end{pmatrix} \times \begin{pmatrix} 2 & 5 \\ -2 & -4 \\ 0 & -1 \end{pmatrix} = \begin{pmatrix} 8 & 18 \\ 18 & 42 \end{pmatrix} = S$$

If the elements of matrix S are multiplied by the scalar $1/(N-1)$ we get the covariance matrix C:

$$C = \frac{1}{N-1} S = \begin{pmatrix} 4 & 9 \\ 9 & 21 \end{pmatrix}$$

The diagonal elements in C contain the variances, the off-diagonal elements the covariances:

$$C = \begin{pmatrix} s_1^2 & r_{12}s_1s_2 \\ r_{12}s_1s_2 & s_2^2 \end{pmatrix}$$

The determinant of C is $|C| = s_1^2 s_2^2 (1 - r_{12}^2)$. This determinant is the multivariate counterpart of the univariate variance. It expresses, in a single number, the amount of spreading or variance in a multivariate distribution.

The matrix D in the definition of the correlation matrix is a diagonal matrix. If we take the square roots of the diagonal elements of matrix S and enter the reciprocals of these square roots in a diagonal matrix, matrix D is obtained:

$$D = \begin{pmatrix} 1/\sqrt{x_1^2} & 0 \\ 0 & 1/\sqrt{x_2^2} \end{pmatrix} = \begin{pmatrix} 1/\sqrt{8} & 0 \\ 0 & 1/\sqrt{42} \end{pmatrix}$$

Matrix R becomes:

$$\begin{aligned} R = DSD &= \begin{pmatrix} 1/\sqrt{8} & 0 \\ 0 & 1/\sqrt{42} \end{pmatrix} \begin{pmatrix} 8 & 18 \\ 18 & 42 \end{pmatrix} \begin{pmatrix} 1/\sqrt{8} & 0 \\ 0 & 1/\sqrt{42} \end{pmatrix} \\ &= \begin{pmatrix} 1.00 & .98 \\ .98 & 1.00 \end{pmatrix} \end{aligned}$$

Appendix C : Statistical tables

Table A: Normal Distribution
Table B: Critical values t
Table C: Critical values F

The probability values in the statistical tables were obtained by using the inverse distribution functions in SPSS.

Table A: Upper percentage points of the Normal Distribution. Example: the one-tailed p of $z \geq 1.15 = .1251$. The p value is found by combining the row for $z = 1.1$ with the column under $z = .05$ ($1.1 + .05 = 1.15$).

z	.00	.01	.02	.03	.04	.05	.06	.07	.08	.09
.0	.5000	.4960	.4920	.4880	.4840	.4801	.4761	.4721	.4681	.4641
.1	.4602	.4562	.4522	.4483	.4443	.4404	.4364	.4325	.4286	.4247
.2	.4207	.4168	.4129	.4090	.4052	.4013	.3974	.3936	.3897	.3859
.3	.3821	.3783	.3745	.3707	.3669	.3632	.3594	.3557	.3520	.3483
.4	.3446	.3409	.3372	.3336	.3300	.3264	.3228	.3192	.3156	.3121
.5	.3085	.3050	.3015	.2981	.2946	.2912	.2877	.2843	.2810	.2776
.6	.2743	.2709	.2676	.2643	.2611	.2578	.2546	.2514	.2483	.2451
.7	.2420	.2389	.2358	.2327	.2296	.2266	.2236	.2206	.2177	.2148
.8	.2119	.2090	.2061	.2033	.2005	.1977	.1949	.1922	.1894	.1867
.9	.1841	.1814	.1788	.1762	.1736	.1711	.1685	.1660	.1635	.1611
1.0	.1587	.1562	.1539	.1515	.1492	.1469	.1446	.1423	.1401	.1379
1.1	.1357	.1335	.1314	.1292	.1271	.1251	.1230	.1210	.1190	.1170
1.2	.1151	.1131	.1112	.1093	.1075	.1056	.1038	.1020	.1003	.0985
1.3	.0968	.0951	.0934	.0918	.0901	.0885	.0869	.0853	.0838	.0823
1.4	.0808	.0793	.0778	.0764	.0749	.0735	.0721	.0708	.0694	.0681
1.5	.0668	.0655	.0643	.0630	.0618	.0606	.0594	.0582	.0571	.0559
1.6	.0548	.0537	.0526	.0516	.0505	.0495	.0485	.0475	.0465	.0455
1.7	.0446	.0436	.0427	.0418	.0409	.0401	.0392	.0384	.0375	.0367
1.8	.0359	.0351	.0344	.0334	.0329	.0322	.0314	.0307	.0301	.0294
1.9	.0287	.0281	.0274	.0268	.0262	.0256	.0250	.0244	.0239	.0233
2.0	.0228	.0222	.0217	.0212	.0207	.0202	.0197	.0192	.0188	.0183
2.1	.0179	.0174	.0170	.0166	.0162	.0158	.0154	.0150	.0146	.0143
2.2	.0139	.0136	.0132	.0129	.0125	.0122	.0119	.0116	.0113	.0110
2.3	.0107	.0104	.0102	.0099	.0096	.0094	.0091	.0089	.0087	.0084
2.4	.0082	.0080	.0078	.0075	.0073	.0071	.0069	.0068	.0066	.0064
2.5	.0062	.0060	.0059	.0057	.0055	.0054	.0052	.0051	.0049	.0048
2.6	.0047	.0045	.0044	.0043	.0041	.0040	.0039	.0038	.0037	.0036
2.7	.0035	.0034	.0033	.0032	.0031	.0030	.0029	.0028	.0027	.0026
2.8	.0026	.0025	.0024	.0023	.0023	.0022	.0021	.0021	.0020	.0019
2.9	.0019	.0018	.0018	.0017	.0016	.0016	.0015	.0015	.0014	.0014
3.0	.0013	.0013	.0013	.0012	.0012	.0011	.0011	.0011	.0010	.0010
3.1	.0010	.0009	.0009	.0009	.0008	.0008	.0008	.0008	.0007	.0007
3.2	.0007									
3.3	.0005									
3.4	.0003									
3.5	.00023									
3.6	.00016									
3.7	.00011									
3.8	.00007									
3.9	.00005									
4.0	.00003									

Table B: Entries in this table give t_α values, where α is the area of the probability in the upper tail of the t distribution. An example: with df is 15 and an α of .025 in the upper tail, $t_{.025} = 2.131$. For two-tailed tests the α level has to be divided by two.

df	.10	.05	.025	.01	.005
1	3.078	6.314	12.706	31.821	63.657
2	1.886	2.920	4.303	6.965	9.925
3	1.683	2.353	3.182	4.541	5.841
4	1.533	2.132	2.776	3.747	4.604
5	1.476	2.015	2.571	3.365	4.032
6	1.440	1.943	2.447	3.143	3.707
7	1.415	1.895	2.365	2.998	3.499
8	1.397	1.860	2.306	2.896	3.355
9	1.383	1.833	2.262	2.821	3.250
10	1.372	1.812	2.228	2.764	3.169
11	1.363	1.796	2.201	2.718	3.106
12	1.356	1.782	2.179	2.681	3.055
13	1.372	1.771	2.160	2.650	3.012
14	1.345	1.761	2.145	2.624	2.977
15	1.341	1.753	2.131	2.602	2.947
16	1.337	1.746	2.120	2.583	2.921
17	1.333	1.740	2.110	2.567	2.898
18	1.330	1.734	2.101	2.552	2.878
19	1.328	1.729	2.093	2.539	2.861
20	1.325	1.725	2.086	2.528	2.845
21	1.323	1.721	2.080	2.518	2.831
22	1.321	1.717	2.074	2.508	2.819
23	1.319	1.714	2.069	2.500	2.807
24	1.318	1.711	2.064	2.492	2.979
25	1.316	1.708	2.060	2.485	2.787
26	1.315	1.706	2.056	2.479	2.779
27	1.314	1.703	2.052	2.473	2.771
28	1.313	1.701	2.048	2.467	2.763
29	1.311	1.699	2.045	2.462	2.756
30	1.310	1.697	2.042	2.457	2.750
40	1.303	1.684	2.021	2.423	2.704
50	1.298	1.675	2.007	2.400	2.674
60	1.296	1.671	2.000	2.390	2.660
80	1.292	1.664	1.990	2.374	2.639
100	1.290	1.660	1.984	2.364	2.626
200	1.286	1.653	1.972	2.345	2.601
500	1.283	1.648	1.965	2.334	2.586
∞	1.282	1.645	1.960	2.326	2.576

Table C

Critical values of F

5% and 1% (bold-face) points for the distribution of F

df_1 = degrees of freedom numerator; df_2 = degrees of freedom denominator

df_2 \ df_1	1	2	3	4	5	6	7	8	9	10	11	12	14	16	20	24	30	40	50	75	100	200	500	∞
1	161	200	216	225	230	234	237	239	241	242	243	244	245	246	248	249	250	251	252	253	253	254	254	254
	4052	**4999**	**5403**	**5625**	**5764**	**5859**	**5928**	**5981**	**6022**	**6056**	**6082**	**6106**	**6142**	**6169**	**6208**	**6234**	**6258**	**6286**	**6302**	**6323**	**6334**	**6352**	**6361**	**6366**
2	18.51	19.00	19.16	19.25	19.30	19.33	19.36	19.37	19.38	19.39	19.40	19.41	19.42	19.43	19.44	19.45	19.46	19.47	19.47	19.48	19.49	19.49	19.50	19.50
	98.49	**99.01**	**99.17**	**99.25**	**99.30**	**99.33**	**99.34**	**99.36**	**99.38**	**99.40**	**99.41**	**99.42**	**99.43**	**99.44**	**99.45**	**99.46**	**99.47**	**99.48**	**99.48**	**99.49**	**99.49**	**99.49**	**99.50**	**99.50**
3	10.13	9.55	9.28	9.12	9.01	8.94	8.88	8.84	8.81	8.78	8.76	8.74	8.71	8.69	8.66	8.64	8.62	8.60	8.58	8.57	8.56	8.54	8.54	8.53
	34.12	**30.81**	**29.46**	**28.71**	**28.24**	**27.91**	**27.67**	**27.49**	**27.34**	**27.23**	**27.13**	**27.05**	**26.92**	**26.83**	**26.69**	**26.60**	**26.50**	**26.41**	**26.35**	**26.27**	**26.23**	**26.18**	**26.14**	**26.12**
4	7.71	6.94	6.59	6.39	6.26	6.16	6.09	6.04	6.00	5.96	5.93	5.91	5.87	5.84	5.80	5.77	5.74	5.71	5.70	5.68	5.66	5.65	5.64	5.63
	21.20	**18.00**	**16.69**	**15.98**	**15.52**	**15.21**	**14.98**	**14.80**	**14.66**	**14.54**	**14.45**	**14.37**	**14.24**	**14.15**	**14.02**	**13.93**	**13.83**	**13.74**	**13.69**	**13.61**	**13.57**	**13.52**	**13.48**	**13.46**
5	6.61	5.79	5.41	5.19	5.05	4.95	4.88	4.82	4.78	4.74	4.70	4.68	4.64	4.60	4.56	4.53	4.50	4.46	4.44	4.42	4.40	4.38	4.37	4.36
	16.26	**13.27**	**12.06**	**11.39**	**10.97**	**10.67**	**10.45**	**10.27**	**10.15**	**10.05**	**9.96**	**9.89**	**9.77**	**9.68**	**9.55**	**9.47**	**9.38**	**9.29**	**9.24**	**9.17**	**9.13**	**9.07**	**9.04**	**9.02**
6	5.99	5.14	4.76	4.53	4.39	4.28	4.21	4.15	4.10	4.06	4.03	4.00	3.96	3.92	3.87	3.84	3.81	3.77	3.75	3.72	3.71	3.69	3.68	3.67
	13.74	**10.92**	**9.78**	**9.15**	**8.75**	**8.47**	**8.26**	**8.10**	**7.98**	**7.87**	**7.79**	**7.72**	**7.60**	**7.52**	**7.39**	**7.31**	**7.23**	**7.14**	**7.09**	**7.02**	**6.99**	**6.94**	**6.90**	**6.88**
7	5.59	4.74	4.35	4.12	3.97	3.87	3.79	3.73	3.68	3.63	3.60	3.57	3.52	3.49	3.44	3.41	3.38	3.34	3.32	3.29	3.28	3.25	3.24	3.23
	12.25	**9.55**	**8.45**	**7.85**	**7.46**	**7.19**	**7.00**	**6.84**	**6.71**	**6.62**	**6.54**	**6.47**	**6.35**	**6.27**	**6.15**	**6.07**	**5.98**	**5.90**	**5.85**	**5.78**	**5.75**	**5.70**	**5.67**	**5.65**
8	5.32	4.46	4.07	3.84	3.69	3.58	3.50	3.44	3.39	3.34	3.31	3.28	3.23	3.20	3.15	3.12	3.08	3.05	3.03	3.00	2.98	2.96	2.94	2.93
	11.26	**8.65**	**7.59**	**7.01**	**6.63**	**6.37**	**6.19**	**6.03**	**5.91**	**5.82**	**5.74**	**5.67**	**5.56**	**5.48**	**5.36**	**5.28**	**5.20**	**5.11**	**5.06**	**5.00**	**4.96**	**4.91**	**4.88**	**4.86**
9	5.12	4.26	3.86	3.63	3.48	3.37	3.29	3.23	3.18	3.13	3.10	3.07	3.02	2.98	2.93	2.90	2.86	2.82	2.80	2.77	2.76	2.73	2.72	2.71
	10.56	**8.02**	**6.99**	**6.42**	**6.06**	**5.80**	**5.62**	**5.47**	**5.35**	**5.26**	**5.18**	**5.11**	**5.00**	**4.92**	**4.80**	**4.73**	**4.64**	**4.56**	**4.51**	**4.45**	**4.41**	**4.36**	**4.33**	**4.31**
10	4.96	4.10	3.71	3.48	3.33	3.22	3.14	3.07	3.02	2.97	2.94	2.91	2.86	2.82	2.77	2.74	2.70	2.67	2.64	2.61	2.59	2.56	2.55	2.54
	10.04	**7.56**	**6.55**	**5.99**	**5.64**	**5.39**	**5.21**	**5.06**	**4.95**	**4.85**	**4.78**	**4.71**	**4.60**	**4.52**	**4.41**	**4.33**	**4.25**	**4.17**	**4.12**	**4.05**	**4.01**	**3.96**	**3.93**	**3.91**
11	4.84	3.98	3.59	3.36	3.20	3.09	3.01	2.95	2.90	2.86	2.82	2.79	2.74	2.70	2.65	2.61	2.57	2.53	2.50	2.47	2.45	2.42	2.41	2.40
	9.65	**7.20**	**6.22**	**5.67**	**5.32**	**5.07**	**4.88**	**4.74**	**4.63**	**4.54**	**4.46**	**4.40**	**4.29**	**4.21**	**4.10**	**4.02**	**3.94**	**3.86**	**3.80**	**3.74**	**3.70**	**3.66**	**3.62**	**3.60**
12	4.75	3.88	3.49	3.26	3.11	3.00	2.92	2.85	2.80	2.76	2.72	2.69	2.64	2.60	2.54	2.50	2.46	2.42	2.40	2.36	2.35	2.32	2.31	2.30
	9.33	**6.93**	**5.95**	**5.41**	**5.06**	**4.82**	**4.65**	**4.50**	**4.39**	**4.30**	**4.22**	**4.16**	**4.05**	**3.98**	**3.86**	**3.78**	**3.70**	**3.61**	**3.56**	**3.49**	**3.46**	**3.41**	**3.38**	**3.36**

Table C: Critical values of F (continued)

df_2	df_1: 1	2	3	4	5	6	7	8	9	10	11	12	14	16	20	24	30	40	50	75	100	200	500	∞
13	4.67	3.80	3.41	3.18	3.02	2.92	2.84	2.77	2.72	2.67	2.63	2.60	2.55	2.51	2.46	2.42	2.38	2.34	2.32	2.28	2.26	2.24	2.22	2.21
	9.07	**6.70**	**5.74**	**5.20**	**4.86**	**4.62**	**4.44**	**4.30**	**4.19**	**4.10**	**4.02**	**3.96**	**3.85**	**3.78**	**3.67**	**3.59**	**3.51**	**3.42**	**3.37**	**3.30**	**3.27**	**3.21**	**3.18**	**3.16**
14	4.60	3.74	3.34	3.11	2.96	2.85	2.77	2.70	2.65	2.60	2.56	2.53	2.48	2.44	2.39	2.35	2.31	2.27	2.24	2.21	2.19	2.16	2.14	2.13
	8.86	**6.51**	**5.56**	**5.03**	**4.69**	**4.46**	**4.28**	**4.14**	**4.03**	**3.94**	**3.86**	**3.80**	**3.70**	**3.62**	**3.51**	**3.43**	**3.34**	**3.26**	**3.21**	**3.14**	**3.11**	**3.06**	**3.02**	**3.00**
15	4.54	3.68	3.29	3.06	2.90	2.79	2.70	2.64	2.59	2.55	2.51	2.48	2.43	2.39	2.33	2.29	2.25	2.21	2.18	2.15	2.12	2.10	2.08	2.07
	8.68	**6.36**	**5.42**	**4.89**	**4.56**	**4.32**	**4.14**	**4.00**	**3.89**	**3.80**	**3.73**	**3.67**	**3.56**	**3.48**	**3.36**	**3.29**	**3.20**	**3.12**	**3.07**	**3.00**	**2.97**	**2.92**	**2.89**	**2.87**
16	4.49	3.63	3.24	3.01	2.85	2.74	2.66	2.59	2.54	2.49	2.45	2.42	2.37	2.33	2.28	2.24	2.20	2.16	2.13	2.09	2.07	2.04	2.02	2.01
	8.53	**6.23**	**5.29**	**4.77**	**4.44**	**4.20**	**4.03**	**3.89**	**3.78**	**3.69**	**3.61**	**3.55**	**3.45**	**3.37**	**3.25**	**3.18**	**3.10**	**3.01**	**2.96**	**2.89**	**2.86**	**2.80**	**2.77**	**2.75**
17	4.45	3.59	3.20	2.96	2.81	2.70	2.62	2.55	2.50	2.45	2.41	2.38	2.33	2.29	2.23	2.19	2.15	2.11	2.08	2.04	2.02	1.99	1.97	1.96
	8.40	**6.11**	**5.18**	**4.67**	**4.34**	**4.10**	**3.93**	**3.79**	**3.68**	**3.59**	**3.52**	**3.45**	**3.35**	**3.27**	**3.16**	**3.08**	**3.00**	**2.92**	**2.86**	**2.79**	**2.76**	**2.70**	**2.67**	**2.65**
18	4.41	3.55	3.16	2.93	2.77	2.66	2.58	2.51	2.46	2.41	2.37	2.34	2.29	2.25	2.19	2.15	2.11	2.07	2.04	2.00	1.98	1.95	1.93	1.92
	8.28	**6.01**	**5.09**	**4.58**	**4.25**	**4.01**	**3.85**	**3.71**	**3.60**	**3.51**	**3.44**	**3.37**	**3.27**	**3.19**	**3.07**	**3.00**	**2.91**	**2.83**	**2.78**	**2.71**	**2.68**	**2.62**	**2.59**	**2.57**
19	4.38	3.52	3.13	2.90	2.74	2.63	2.55	2.48	2.43	2.38	2.34	2.31	2.26	2.21	2.15	2.11	2.07	2.02	2.00	1.96	1.94	1.91	1.90	1.88
	8.18	**5.93**	**5.01**	**4.50**	**4.17**	**3.94**	**3.77**	**3.63**	**3.52**	**3.43**	**3.36**	**3.30**	**3.19**	**3.12**	**3.00**	**2.92**	**2.84**	**2.76**	**2.70**	**2.63**	**2.60**	**2.54**	**2.51**	**2.49**
20	4.35	3.49	3.10	2.87	2.71	2.60	2.52	2.45	2.40	2.35	2.31	2.28	2.23	2.18	2.12	2.08	2.04	1.99	1.96	1.92	1.90	1.87	1.85	1.84
	8.10	**5.85**	**4.94**	**4.43**	**4.10**	**3.87**	**3.71**	**3.56**	**3.45**	**3.37**	**3.30**	**3.23**	**3.13**	**3.05**	**2.94**	**2.86**	**2.77**	**2.69**	**2.63**	**2.56**	**2.53**	**2.47**	**2.44**	**2.42**
21	4.32	3.47	3.07	2.84	2.68	2.57	2.49	2.42	2.37	2.32	2.28	2.25	2.20	2.15	2.09	2.05	2.00	1.96	1.93	1.89	1.87	1.84	1.82	1.81
	8.02	**5.78**	**4.87**	**4.37**	**4.04**	**3.81**	**3.65**	**3.51**	**3.40**	**3.31**	**3.24**	**3.17**	**3.07**	**2.99**	**2.88**	**2.80**	**2.72**	**2.63**	**2.58**	**2.51**	**2.47**	**2.42**	**2.38**	**2.36**
22	4.30	3.44	3.05	2.82	2.66	2.55	2.47	2.40	2.35	2.30	2.26	2.23	2.18	2.13	2.07	2.03	1.98	1.93	1.91	1.87	1.84	1.81	1.80	1.78
	7.94	**5.72**	**4.82**	**4.31**	**3.99**	**3.76**	**3.59**	**3.45**	**3.35**	**3.26**	**3.18**	**3.12**	**3.02**	**2.94**	**2.83**	**2.75**	**2.67**	**2.58**	**2.53**	**2.46**	**2.42**	**2.37**	**2.33**	**2.31**
23	4.28	3.42	3.03	2.80	2.64	2.53	2.45	2.38	2.32	2.28	2.24	2.20	2.14	2.10	2.04	2.00	1.96	1.91	1.88	1.84	1.82	1.79	1.77	1.76
	7.88	**5.66**	**4.76**	**4.26**	**3.94**	**3.71**	**3.54**	**3.41**	**3.30**	**3.21**	**3.14**	**3.07**	**2.97**	**2.89**	**2.78**	**2.70**	**2.62**	**2.53**	**2.48**	**2.41**	**2.37**	**2.32**	**2.28**	**2.26**
24	4.26	3.40	3.01	2.78	2.62	2.51	2.43	2.36	2.30	2.26	2.22	2.18	2.13	2.09	2.02	1.98	1.94	1.89	1.86	1.82	1.80	1.76	1.74	1.73
	7.82	**5.61**	**4.72**	**4.22**	**3.90**	**3.67**	**3.50**	**3.36**	**3.25**	**3.17**	**3.09**	**3.03**	**2.93**	**2.85**	**2.74**	**2.66**	**2.58**	**2.49**	**2.44**	**2.36**	**2.33**	**2.27**	**2.23**	**2.21**
25	4.24	3.38	2.99	2.76	2.60	2.49	2.41	2.34	2.28	2.24	2.20	2.16	2.11	2.06	2.00	1.96	1.92	1.87	1.84	1.80	1.77	1.74	1.72	1.71
	7.77	**5.57**	**4.68**	**4.18**	**3.86**	**3.63**	**3.46**	**3.32**	**3.21**	**3.13**	**3.05**	**2.99**	**2.89**	**2.81**	**2.70**	**2.62**	**2.54**	**2.45**	**2.40**	**2.32**	**2.29**	**2.23**	**2.19**	**2.17**

Table C: Critical values of *F* (continued)

df_2	n_1: 1	2	3	4	5	6	7	8	9	10	11	12	14	16	20	24	30	40	50	75	100	200	500	∞
26	4.22	3.37	2.98	2.74	2.59	2.47	2.39	2.32	2.27	2.22	2.18	2.15	2.10	2.05	1.99	1.95	1.90	1.85	1.82	1.78	1.76	1.72	1.70	1.69
	7.72	**5.53**	**4.64**	**4.14**	**3.82**	**3.59**	**3.42**	**3.29**	**3.17**	**3.09**	**3.02**	**2.96**	**2.86**	**2.77**	**2.66**	**2.58**	**2.50**	**2.41**	**2.36**	**2.28**	**2.25**	**2.19**	**2.15**	**2.13**
27	4.21	3.35	2.96	2.73	2.57	2.46	2.37	2.30	2.25	2.20	2.16	2.13	2.08	2.03	1.97	1.93	1.88	1.84	1.80	1.76	1.74	1.71	1.68	1.67
	7.68	**5.49**	**4.60**	**4.11**	**3.79**	**3.56**	**3.39**	**3.26**	**3.14**	**3.06**	**2.98**	**2.93**	**2.83**	**2.74**	**2.63**	**2.55**	**2.47**	**2.38**	**2.33**	**2.25**	**2.21**	**2.16**	**2.12**	**2.10**
28	4.20	3.34	2.95	2.71	2.56	2.44	2.36	2.29	2.24	2.19	2.15	2.12	2.06	2.02	1.96	1.91	1.87	1.81	1.78	1.75	1.72	1.69	1.67	1.65
	7.64	**5.45**	**4.57**	**4.07**	**3.76**	**3.53**	**3.36**	**3.23**	**3.11**	**3.03**	**2.95**	**2.90**	**2.80**	**2.71**	**2.60**	**2.52**	**2.44**	**2.35**	**2.30**	**2.22**	**2.18**	**2.13**	**2.09**	**2.06**
29	4.18	3.33	2.93	2.70	2.54	2.43	2.35	2.28	2.22	2.18	2.14	2.10	2.05	2.00	1.94	1.90	1.85	1.80	1.77	1.73	1.71	1.68	1.65	1.64
	7.60	**5.42**	**4.54**	**4.04**	**3.73**	**3.50**	**3.33**	**3.20**	**3.08**	**3.00**	**2.92**	**2.87**	**2.77**	**2.68**	**2.57**	**2.49**	**2.41**	**2.32**	**2.27**	**2.19**	**2.15**	**2.10**	**2.06**	**2.03**
30	4.17	3.32	2.92	2.69	2.53	2.42	2.34	2.27	2.21	2.16	2.12	2.09	2.04	1.99	1.93	1.89	1.84	1.79	1.76	1.72	1.69	1.66	1.64	1.62
	7.56	**5.39**	**4.51**	**4.02**	**3.70**	**3.47**	**3.30**	**3.17**	**3.06**	**2.98**	**2.90**	**2.84**	**2.74**	**2.66**	**2.55**	**2.47**	**2.38**	**2.29**	**2.24**	**2.16**	**2.13**	**2.07**	**2.03**	**2.01**
32	4.15	3.30	2.90	2.67	2.51	2.40	2.32	2.25	2.19	2.14	2.10	2.07	2.02	1.97	1.91	1.86	1.82	1.76	1.74	1.69	1.67	1.64	1.61	1.59
	7.50	**5.34**	**4.46**	**3.97**	**3.66**	**3.42**	**3.25**	**3.12**	**3.01**	**2.94**	**2.86**	**2.80**	**2.70**	**2.62**	**2.51**	**2.42**	**2.34**	**2.25**	**2.20**	**2.12**	**2.08**	**2.02**	**1.98**	**1.96**
34	4.13	3.28	2.88	2.65	2.49	2.38	2.30	2.23	2.17	2.12	2.08	2.05	2.00	1.95	1.89	1.84	1.80	1.74	1.71	1.67	1.64	1.61	1.59	1.57
	7.44	**5.29**	**4.42**	**3.93**	**3.61**	**3.38**	**3.21**	**3.08**	**2.97**	**2.89**	**2.82**	**2.76**	**2.66**	**2.58**	**2.47**	**2.38**	**2.30**	**2.21**	**2.15**	**2.08**	**2.04**	**1.98**	**1.94**	**1.91**
36	4.11	3.26	2.86	2.63	2.48	2.36	2.28	2.21	2.15	2.10	2.06	2.03	1.98	1.93	1.87	1.82	1.78	1.72	1.69	1.65	1.62	1.59	1.56	1.55
	7.39	**5.25**	**4.38**	**3.89**	**3.58**	**3.35**	**3.18**	**3.04**	**2.94**	**2.86**	**2.78**	**2.72**	**2.62**	**2.54**	**2.43**	**2.35**	**2.26**	**2.17**	**2.12**	**2.04**	**2.00**	**1.94**	**1.90**	**1.87**
38	4.10	3.25	2.85	2.62	2.46	2.35	2.26	2.19	2.14	2.09	2.05	2.02	1.96	1.92	1.85	1.80	1.76	1.71	1.67	1.63	1.60	1.57	1.54	1.53
	7.35	**5.21**	**4.34**	**3.86**	**3.54**	**3.32**	**3.15**	**3.02**	**2.91**	**2.82**	**2.75**	**2.69**	**2.59**	**2.51**	**2.40**	**2.32**	**2.22**	**2.14**	**2.08**	**2.00**	**1.97**	**1.90**	**1.86**	**1.84**
40	4.08	3.23	2.84	2.61	2.45	2.34	2.25	2.18	2.12	2.07	2.04	2.00	1.95	1.90	1.84	1.79	1.74	1.69	1.66	1.61	1.59	1.55	1.53	1.51
	7.31	**5.18**	**4.31**	**3.83**	**3.51**	**3.29**	**3.12**	**2.99**	**2.88**	**2.80**	**2.73**	**2.66**	**2.56**	**2.49**	**2.37**	**2.29**	**2.20**	**2.11**	**2.05**	**1.97**	**1.94**	**1.88**	**1.84**	**1.81**
42	4.07	3.22	2.83	2.59	2.44	2.32	2.24	2.17	2.11	2.06	2.02	1.99	1.94	1.89	1.82	1.78	1.73	1.68	1.64	1.60	1.57	1.54	1.51	1.49
	7.27	**5.15**	**4.29**	**3.80**	**3.49**	**3.26**	**3.10**	**2.96**	**2.86**	**2.77**	**2.70**	**2.64**	**2.54**	**2.46**	**2.35**	**2.26**	**2.17**	**2.08**	**2.02**	**1.94**	**1.91**	**1.85**	**1.80**	**1.78**
44	4.06	3.21	2.82	2.58	2.43	2.31	2.23	2.16	2.10	2.05	2.01	1.98	1.92	1.88	1.81	1.76	1.72	1.66	1.63	1.58	1.56	1.52	1.50	1.48
	7.24	**5.12**	**4.26**	**3.78**	**3.46**	**3.24**	**3.07**	**2.94**	**2.84**	**2.75**	**2.68**	**2.62**	**2.52**	**2.44**	**2.32**	**2.24**	**2.15**	**2.06**	**2.00**	**1.92**	**1.88**	**1.82**	**1.78**	**1.75**
46	4.05	3.20	2.81	2.57	2.42	2.30	2.22	2.14	2.09	2.04	2.00	1.97	1.91	1.87	1.80	1.75	1.71	1.65	1.62	1.57	1.54	1.51	1.48	1.46
	7.21	**5.10**	**4.24**	**3.76**	**3.44**	**3.22**	**3.05**	**2.92**	**2.82**	**2.73**	**2.66**	**2.60**	**2.50**	**2.42**	**2.30**	**2.22**	**2.13**	**2.04**	**1.98**	**1.90**	**1.86**	**1.80**	**1.76**	**1.72**
48	4.04	3.19	2.80	2.56	2.41	2.30	2.21	2.14	2.08	2.03	1.99	1.96	1.90	1.86	1.79	1.74	1.70	1.64	1.61	1.56	1.53	1.50	1.47	1.45
	7.19	**5.08**	**4.22**	**3.74**	**3.42**	**3.20**	**3.04**	**2.90**	**2.80**	**2.71**	**2.64**	**2.58**	**2.48**	**2.40**	**2.28**	**2.20**	**2.11**	**2.02**	**1.96**	**1.88**	**1.84**	**1.78**	**1.73**	**1.70**

Table C: Critical values of F (continued)

df_2	n_1 = 1	2	3	4	5	6	7	8	9	10	11	12	14	16	20	24	30	40	50	75	100	200	500	∞
50	4.03	3.18	2.79	2.56	2.40	2.29	2.20	2.13	2.07	2.02	1.98	1.95	1.90	1.85	1.78	1.74	1.69	1.63	1.60	1.55	1.52	1.48	1.46	1.44
	7.17	**5.06**	**4.20**	**3.72**	**3.41**	**3.18**	**3.02**	**2.88**	**2.78**	**2.70**	**2.62**	**2.56**	**2.46**	**2.39**	**2.26**	**2.18**	**2.10**	**2.00**	**1.94**	**1.86**	**1.82**	**1.76**	**1.71**	**1.68**
55	4.02	3.17	2.78	2.54	2.38	2.27	2.18	2.11	2.05	2.00	1.97	1.93	1.88	1.83	1.76	1.72	1.67	1.61	1.58	1.52	1.50	1.46	1.43	1.41
	7.12	**5.01**	**4.16**	**3.68**	**3.37**	**3.15**	**2.98**	**2.85**	**2.75**	**2.66**	**2.59**	**2.53**	**2.43**	**2.35**	**2.23**	**2.15**	**2.06**	**1.96**	**1.90**	**1.82**	**1.78**	**1.71**	**1.66**	**1.64**
60	4.00	3.15	2.76	2.52	2.37	2.25	2.17	2.10	2.04	1.99	1.95	1.92	1.86	1.81	1.75	1.70	1.65	1.59	1.56	1.50	1.48	1.44	1.41	1.39
	7.08	**4.98**	**4.13**	**3.65**	**3.34**	**3.12**	**2.95**	**2.82**	**2.72**	**2.63**	**2.56**	**2.50**	**2.40**	**2.32**	**2.20**	**2.12**	**2.03**	**1.93**	**1.87**	**1.79**	**1.74**	**1.68**	**1.63**	**1.60**
65	3.99	3.14	2.75	2.51	2.36	2.24	2.15	2.08	2.02	1.98	1.94	1.90	1.85	1.80	1.73	1.68	1.63	1.57	1.54	1.49	1.46	1.42	1.39	1.37
	7.04	**4.95**	**4.10**	**3.62**	**3.31**	**3.09**	**2.93**	**2.79**	**2.70**	**2.61**	**2.54**	**2.47**	**2.37**	**2.30**	**2.18**	**2.09**	**2.00**	**1.90**	**1.84**	**1.76**	**1.71**	**1.64**	**1.60**	**1.56**
70	3.98	3.13	2.74	2.50	2.35	2.23	2.14	2.07	2.01	1.97	1.93	1.89	1.84	1.79	1.72	1.67	1.62	1.56	1.53	1.47	1.45	1.40	1.37	1.35
	7.01	**4.92**	**4.08**	**3.60**	**3.29**	**3.07**	**2.91**	**2.77**	**2.67**	**2.59**	**2.51**	**2.45**	**2.35**	**2.28**	**2.15**	**2.07**	**1.98**	**1.88**	**1.82**	**1.74**	**1.69**	**1.62**	**1.56**	**1.53**
80	3.96	3.11	2.72	2.48	2.33	2.21	2.12	2.05	1.99	1.95	1.91	1.88	1.82	1.77	1.70	1.65	1.60	1.54	1.51	1.45	1.42	1.38	1.35	1.32
	6.96	**4.88**	**4.04**	**3.56**	**3.25**	**3.04**	**2.87**	**2.74**	**2.64**	**2.55**	**2.48**	**2.41**	**2.32**	**2.24**	**2.11**	**2.03**	**1.94**	**1.84**	**1.78**	**1.70**	**1.65**	**1.57**	**1.52**	**1.49**
100	3.94	3.09	2.70	2.46	2.30	2.19	2.10	2.03	1.97	1.92	1.88	1.85	1.79	1.75	1.68	1.63	1.57	1.51	1.48	1.42	1.39	1.34	1.30	1.28
	6.90	**4.82**	**3.98**	**3.51**	**3.20**	**2.99**	**2.82**	**2.69**	**2.59**	**2.51**	**2.43**	**2.36**	**2.26**	**2.19**	**2.06**	**1.98**	**1.89**	**1.79**	**1.73**	**1.64**	**1.59**	**1.51**	**1.46**	**1.43**
125	3.92	3.07	2.68	2.44	2.29	2.17	2.08	2.01	1.95	1.90	1.86	1.83	1.77	1.72	1.65	1.60	1.55	1.49	1.45	1.39	1.36	1.31	1.27	1.25
	6.84	**4.78**	**3.94**	**3.47**	**3.17**	**2.95**	**2.79**	**2.65**	**2.56**	**2.47**	**2.40**	**2.33**	**2.23**	**2.15**	**2.03**	**1.94**	**1.85**	**1.75**	**1.68**	**1.59**	**1.54**	**1.46**	**1.40**	**1.37**
150	3.91	3.06	2.67	2.43	2.27	2.16	2.07	2.00	1.94	1.89	1.85	1.82	1.76	1.71	1.64	1.59	1.54	1.47	1.44	1.37	1.34	1.29	1.25	1.22
	6.81	**4.75**	**3.91**	**3.44**	**3.14**	**2.92**	**2.76**	**2.62**	**2.53**	**2.44**	**2.37**	**2.30**	**2.20**	**2.12**	**2.00**	**1.91**	**1.83**	**1.72**	**1.66**	**1.56**	**1.51**	**1.43**	**1.37**	**1.33**
200	3.89	3.04	2.65	2.41	2.26	2.14	2.05	1.98	1.92	1.87	1.83	1.80	1.74	1.69	1.62	1.57	1.52	1.45	1.42	1.35	1.32	1.26	1.22	1.19
	6.76	**4.71**	**3.88**	**3.41**	**3.11**	**2.90**	**2.73**	**2.60**	**2.50**	**2.41**	**2.34**	**2.28**	**2.17**	**2.09**	**1.97**	**1.88**	**1.79**	**1.69**	**1.62**	**1.53**	**1.48**	**1.39**	**1.33**	**1.28**
400	3.86	3.02	2.62	2.39	2.23	2.12	2.03	1.96	1.90	1.85	1.81	1.78	1.72	1.67	1.60	1.54	1.49	1.42	1.38	1.32	1.28	1.22	1.16	1.13
	6.70	**4.66**	**3.83**	**3.36**	**3.06**	**2.85**	**2.69**	**2.55**	**2.46**	**2.37**	**2.29**	**2.23**	**2.12**	**2.04**	**1.92**	**1.84**	**1.74**	**1.64**	**1.57**	**1.47**	**1.42**	**1.32**	**1.24**	**1.19**
1000	3.85	3.00	2.61	2.38	2.22	2.10	2.02	1.95	1.89	1.84	1.80	1.76	1.70	1.65	1.58	1.53	1.47	1.41	1.36	1.30	1.26	1.19	1.13	1.08
	6.66	**4.62**	**3.80**	**3.34**	**3.04**	**2.82**	**2.66**	**2.53**	**2.43**	**2.34**	**2.26**	**2.20**	**2.09**	**2.01**	**1.89**	**1.81**	**1.71**	**1.61**	**1.54**	**1.44**	**1.38**	**1.28**	**1.19**	**1.11**
∞	3.84	2.99	2.60	2.37	2.21	2.09	2.01	1.94	1.88	1.83	1.79	1.75	1.69	1.64	1.57	1.52	1.46	1.40	1.35	1.28	1.24	1.17	1.11	1.00
	6.64	**4.60**	**3.78**	**3.32**	**3.02**	**2.80**	**2.64**	**2.51**	**2.41**	**2.32**	**2.24**	**2.18**	**2.07**	**1.99**	**1.87**	**1.79**	**1.69**	**1.59**	**1.52**	**1.41**	**1.36**	**1.25**	**1.15**	**1.00**

Index

Bei Fragen zur Produktsicherheit wenden Sie sich bitte an:
If you have any questions regarding product safety,
please contact:

Walter de Gruyter GmbH
Genthiner Straße 13
10785 Berlin
productsafety@degruyterbrill.com